Composing Music with Computers

TECHNOLOGY
S e r i e s

Titles in the series

Composing Music with Computers

Eduardo Reck Miranda

AMSTERDAM • BOSTON • HEIDELBERG • LONDON • NEW YORK • OXFORD
PARIS • SAN DIEGO • SAN FRANCISCO • SINGAPORE • SYDNEY • TOKYO

Focal Press is an imprint of Elsevier

ELSEVIER

Focal
Press

Focal Press
An imprint of Elsevier
Linacre House, Jordan Hill, Oxford OX2 8DP
200 Wheeler Road, Burlington, MA 01803

First published 2002
Reprinted 2003, 2004

Permissions may be sought directly from Elsevier's Science and Technology
Rights Department in Oxford, UK: phone: (+44) (0) 1865 843830; fax: (+44) (0)
1865 853333; e-mail: permissions@elsevier.co.uk. You may also complete
your request on-line via the Elsevier homepage (http://www.elsevier.com), by
selecting 'Customer Support' and then 'Obtaining Permissions'

British Library Cataloguing in Publication Data
A catalogue record for this book is available from the British Library

Library of Congress Cataloging in Publication Data
A catalog record for this book is available from the Library of Congress

ISBN 0 240 51567 6

For information on all Focal Press publications
visit our website at www.focalpress.com

Printed and bound in Great Britain by The Bath Press, Bath

Contents

Series introduction

The Focal Press Music Technology Series is intended to fill a growing need for authoritative books to support college and university courses in music technology, sound recording, multimedia and their related fields. The books will also be of value to professionals already working in these areas and who want either to update their knowledge or to familiarise themselves with topics that have not been part of their mainstream occupations.

Information technology and digital systems are now widely used in the production of sound and in the composition of music for a wide range of end uses. Those working in these fields need to understand the principles of sound, musical acoustics, sound synthesis, digital audio, video and computer systems. This is a tall order, but people with this breadth of knowledge are increasingly sought after by employers. The series will explain the technology and techniques in a manner which is both readable and factually concise, avoiding the chattiness, informality and technical woolliness of many books on music technology. The authors are all experts in their fields and many come from teaching and research backgrounds.

Dr Francis Rumsey
Series Editor

Foreword

Composing music with computers is gradually taking its place alongside more traditional ways, such as pencil and paper or improvisation. Computer music is now finding its way into many musical genres, including pop, rock, techno, disco, jazz, and music for film. It is no longer confined to the aesthetics that emerged in the 1960s from the experimental works of *electronic music* (i.e., music composed using electronically synthesised sounds) and *musique concrète* (i.e., music composed using recorded acoustic sound, or sampling music). Gone are the days where computer-aided composition could only be carried out in select institutions using equipment far too expensive for any individual to own. The soundcard, now a standard component in most home computers, outperforms systems that were state-of-the-art a mere 10–15 years ago. Current processor speeds enable real-time synthesis and sound processing even on laptop computers. Consumer music software facilitates musical composition through relatively easy-to-use graphical user-interfaces that require no programming skills. What more could one want?

Electroacoustic music (music that combines *electronic* with *concrète* practices) in its first few decades inspired a somewhat romantic belief that the new technology, by enabling new sounds and compositional process, would in turn lead to a new music with new concepts, aesthetics, and musical experiences. It is not clear whether this has actually occurred, and if so, to what extent. It seems that the second half of the twentieth century, particularly with the move from using analogue devices to digital systems, was occupied largely with problems of 'computation technology', i.e., how to compute music samples fast enough and with good audio quality. There is yet much to be learned about how the computer can help us to express and capture our musical ideas, to experiment with and develop them, and most importantly, to organise them and produce the finished work of art. This problem could be summarised in much simpler words: how to *compose* music with computers.

Consumer applications for musical composition still leave much to be desired. Sequencers, which are so popular for musical composition, model a multi-track recorder that is familiar from the recording studio. This familiarity makes them easy to use but provides little support for the varied musical concepts through which the composer may be conceiving his work. In addition this model emphasises the placing of musical content in time but doesn't directly support generation of musical materials.

Music is a complex and abstract domain and is inherently subjective. This subjectivity expresses itself in a multitude of different concepts and approaches. Traditional music theory, which developed through vocal and instrumental music over several centuries, has provided many concepts that form the building blocks of musical thought. Note and melody to phrase, motif, development, and structure; tension, and relaxation, to voice leading, counterpoint, harmony, and form are but a few. Newer musical forms introduced yet other ways of looking at music. The electroacoustic music of the 1960s and 1970s, made possible for the first time by analogue devices and digital computers, encouraged composers to think differently not only about music and the process of composition but also about the sensation of the musical experience. Electronic music built upon concepts derived from *music serialism* and produced musical complexity beyond the ability of human performance. On the other extreme, *musique concrète* led composers to explore timbre, textures, time flow, transitions, phase shifts, and sound morphs. In short, it is clear that different people think about their music using different concepts. Some of these concepts are more general and to some degree universal, but many are highly individualised and apply only to the musical world of a specific composition.

The important contribution of this book lies in its in-depth survey of the varied approaches and techniques that have been developed and utilised for generating music with computers in the academic community. Each technique represents a different way of thinking about music. With a little imagination the reader may find that a certain technique can map nicely to his unique way of thinking about music. The tools described in this book can be used not only to express and generate new musical ideas but also to process previously composed musical materials. More importantly, the reader can experiment hands-on with each technique through a collection of music applications on the accompanying CD-ROM that have been developed by leading researchers in the field.

The reader will benefit if he keeps in mind several fundamental questions that Eduardo Reck Miranda raises. Is there a difference between instrumental and computer music? Is there a difference in the way one composes a work for instruments or for computer? Is there a different creative workflow when using pencil and paper, the computer, or improvisational techniques? These questions are especially relevant today where computers equally support the composition of traditional and experimental forms of music. The reader may not find a single decisive answer to these questions, but more aptly, many different answers depending on the person who is asking and the specific musical problem he is considering. This book is applicable whether the reader thinks about his music using traditional concepts or through individualised concepts that apply only in his own particular musical world. This book is also relevant whether the reader is concerned with manipulating the overall structure of his work or with refining minute details. It is highly likely that anyone interested in using computers to compose music will find herein a useful technique, regardless of his style or method of composition. In the tradition of computer music – experiment: try, listen, and refine.

Daniel V. Oppenheim
Computer Music Center
IBM T. J. Watson Research Center
Yorktown Heights, NY 10598
USA

Preface

Composers, perhaps more than any other class of artists, have always been acutely aware of the scientific developments of their time. From the discovery, almost three thousand years ago, of the direct relationship between the pitch of a note and the length of a string or pipe, to the latest computer models of human musical cognition and intelligence, composers have always looked to science to provide new and challenging ways to study and compose music.

Music is generally associated with the artistic expression of emotions, but it is clear that reason plays an important role in music making. For example, the ability to recognise musical patterns and to make structural abstractions and associations requires sophisticated memory mechanisms, involving the conscious manipulation of concepts and subconscious access to millions of networked neurological bonds. One of the finest examples of early rational approaches to music composition appeared in the eleventh century in Europe, when Guido d'Arezzo proposed a lookup chart for assigning pitch to the syllables of religious hymns. He also invented the musical stave for systematic notation of music and established the medieval music scales known as the church modes.

Between d'Arezzo's charts and the first compositional computer programs that appeared in the early 1950s, countless systematisations of music for composition purposes were proposed. The use of the computer as a composition tool thus continues the tradition of Western musical thought that was initiated approximately a thousand years ago. The computer is a powerful tool for the realisation of abstract design constructs, enabling composers to create musical systematisations and judge whether they have the potential to produce interesting music.

One of the first hints of machine composition came from Ada Lovelace around 1840 in England. The mathematician Charles Babbage had just conceived his analytical engine (Swade, 1991), a programmable calculating engine now considered to be the precursor of the modern-day computer, and Lovelace proposed the idea of using Babbage's invention to compose pieces of music; unfortunately her ideas were not put into practice at the time. All the same, at around that time steam powered machines controlled by stacks of punched cards were being engineered for the textile industry; musical instrument builders promptly recognised that punch-card stacks could be used to drive automatic pipe organs. This automatic organ revealed a glimpse of an unsettling idea about the nature of the music it produced: it emanated from a stream of information, or data, in a form that might also be used to weave a textile pattern. The idea soon evolved into mechanical pianos (popularly known as 'pianolas') and several companies began as early as the 1900s to manufacture the so-called *reproducing pianos*. Reproducing pianos could generate up to sixteen different shades of loudness and some could vary the volume on a continuous sliding scale. Such pianos enabled pianists to record their work with great fidelity: the recording apparatus could punch four thousands holes per minute on a piano roll, enough to store all the notes that a fast virtuoso could play. Because a piano roll stored a set of parameters that described the sound and not the sounds themselves, the performance remained malleable: the information could be manually edited, the holes re-cut, and so on. This sort of technology gained much sophistication during the course of the twentieth century, but the underlying principles remained the same. The medium has changed from punched cards and paper rolls to magnetic and optical storage, but perhaps the most substantial improvement is the invention of a processor that can perform operations on the stored data as well as create it: *the computer*.

The use of computers as composition generators was pioneered in the mid-1950s by such people as Lejaren Hiller and Leonard Isaacson in the United States of America, whose 1956 work, *The Illiac Suite for String Quartet*, is recognised as being the first computer-composed composition (Manning, 1985). Today, computers play key roles in several aspects of music making, ranging from the synthesis of complex sounds which are impossible to produce with acoustic musical instruments, to the automatic generation of music. The role of the computer as a synthesiser is extensively discussed in my previous book in the Focal Press Music Technology Series, *Computer Sound Synthesis for the Electronic Musician*.

Composing Music with Computers focuses on the role of the computer as a tool for music composition with an emphasis on automatic generation. The first chapter introduces some fundamental concepts concerning compositional approaches and paradigms. Here we will touch upon issues such as music representation, modelling and compositional archetypes. Whilst astronomy, numerical proportions and esoteric numbers served to scaffold some musical pieces in the past, discrete mathematics, set theory, logic, formal grammars, probabilities and algorithms are the mathematical tools of the contemporary composer. In contrast to the discursive style of Chapter 1, Chapter 2 focuses on the mathematical background that musicians should master in order to fully explore the potential of the computer for composition; it is important to study this chapter to take full advantage of this book. Next, Chapter 3 presents the role that probabilities, formal grammars and automata can play in composition, followed by an introduction to the art of composing music with iterative processes, such as fractals, in Chapter 4.

Besides the fact that computers can process enormous amounts of data with precision and considerable speed, computers can also be programmed to mimic some aspects of human cognition and intelligence. One of the most exciting branches of computer science is *neural computation*, whereby scientists study the human brain by analysing its activities when we perform specific tasks, and building computer models that emulate its functioning. Very interesting technology for music making is emerging from this research, such as brain interfaces for music control, and systems that can autonomously learn to recognise and produce musical material. This technology is discussed in Chapter 5.

Compositional techniques involving mathematical models such as the ones introduced in Chapters 3 and 4 have contributed enormously to the radical changes that occurred in Western music after the second half of the twentieth century, leading to the countless tendencies and approaches to musical composition that are recognised today. An emerging new compositional trend that can be considered to be a natural progression in computer music research is presented in Chapter 6 'Evolutionary music'. Here the tools for composition are drawn from research into the origins and evolution of biological organisms, ecology and cultural systems. Next, Chapter 7 presents three case studies on putting some of the techniques and concepts introduced in this book into practice. Finally, Chapter 8 presents the software on the accompanying CD-ROM. The CD-ROM contains examples, complementary tutorials and a number of composition systems mostly for PC and Macintosh platforms, ranging from experimental programs and demonstration versions, to fully working packages developed by research centres, individuals and companies worldwide.

I would like to express my gratitude to all the contributors who kindly provided the materials for the CD-ROM: Roger Dannenberg and Dominic Mazzoni (Carnegie Mellon University, USA); Gerard Assayag and Carlos Agon (Ircam, France); Steve Abrams, Robert Fuhrer, Daniel Oppenheim, Donald Pazel, and James Wright (IBM T. J. Watson Research Center, USA); Tim Cole and Pete Cole (SSEYO Ltd, UK); Bernard Bel (Université de Provence, France); Arnold Reinders (MuSoft Builders, The Netherlands); Kenny McAlpine and Michael McClarty (University of Abertay Dundee, UK); Jônatas Manzolli and Artemis Moroni (Unicamp, Brazil); Valerio Talarico, Eleonora Bilotta and Pietro Pantano (Università della Calabria, Italy); Drew DeVito (IBVA Systems, USA); Sylvia Pengilly (Loyola University, New Orleans, USA); Lars Kindermann (The Bavarian Research Centre for Knowledge-Based Systems, Germany); David Zicarelli (Cycling74, USA); Luis Rojas; Gustavo Diaz and Paul Whalley. I am indebted to Francis Rumsey (University of Surrey, UK), Stuart Hoggar (University of Glasgow, UK), Alan Smaill (University of Edinburgh, UK), Daniel Oppenheim and Peter Todd (Max Planck Institute for Human Development, Germany) who read the manuscript draft and made invaluable comments and suggestions; to James Correa for his help with the musical figures; to Larry Solomon (University of Arizona, USA) for the solution of the puzzle canon in Chapter 4; to Gerald Bennett (Swiss Center for Computer Music, Switzerland) for supplying the annotated score for Appendix 1; to Steve Goss (University of Surrey, UK) for testing the software on the CD-ROM; and to the Focal Press team for their support and professionalism.

Finally I would like to thank my wife, Alexandra, for her patience and for occasionally making me tea.

This book is dedicated to Felipe Brum Reck for his academic inspiration.

Eduardo Reck Miranda

1 Computer music: facing the facts

Perhaps one of the most significant aspects that differentiates humans from other animals is the fact that we are inherently musical. Our compulsion to listen to and appreciate sound arrangements beyond the mere purposes of linguistic communication is extraordinary, and it is indeed this intriguing aspect of humanity that inspires two fundamental notions in this book. The first is the notion that *music is sounds organised in space and time*. In this context, space is primarily associated with vertical (or simultaneous) relationships between sounds, whereas time is associated with horizontal (or sequential) relationships. The notion of space in terms of the geographical distribution of sounds in the performance area is an exciting new dynamic. Contemporary composers are increasingly exploring 'real' (that is, geographical) space in their pieces by either distributing performers at different locations in a room and/or using sophisticated sound diffusion systems.

Our ability to create and appreciate the organisation of sounds in space and time leads us to our second fundamental notion: the notion that *musical compositions carry abstract structures*. These notions are important for computer music because they address issues that are at the fringe of two apparently distinct domains of our musical intelligence: the domain of abstract subjectivity (musical composition and artistic imagination) and the domain of abstract objectivity (logical operations and mathematical thought). There is no dispute that the computer is a tool of excellence for the latter domain, but this book is also interested in exploring the potential of the computer for the former domain.

This chapter presents an introductory discussion on several varied topics of interest concerning compositional approaches and paradigms, with the intention of warming up the engines for the chapters to follow. This discussion is by no means exhaustive: it uncovers only the tip of a giant iceberg immersed in a dense ocean of theories, aesthetic preferences, ideologies, viewpoints and quarrels.

In order to avoid the use of pleonasms such as 'European music', 'Western music', and the like, from now on, unless otherwise stated, the term 'music' in this book refers to music in the Western tradition.

1.1 Abstraction boundaries

The notion that musical compositions carry abstract structures implies the notion of abstraction boundaries. The definition of abstraction boundaries is of primary importance for composers working with computers because it determines the building blocks, or components, that form the musical structures of a composition.

As a starting point for discussion, let us establish three main levels of abstraction: the *microscopic level*, the *note level* and the *building-block level*. At the microscopic level, the composer works with microscopic sound features, such as frequencies and amplitudes of the individual partials of the sounds; here composers normally have access to fine timing controls, even in milliseconds. In this case, a musical score is more likely to contain lists of numerical values for digital sound synthesis instruments, rather than notes for performance on acoustic instruments. The following example is a score devised by Aluízio Arcela written for a software synthesiser in a musical programming language developed at the University of Brasília, called SOM-A:

```
(VAL 0 6 44100)
    (INS 6 instr1
        (1 0 ((0 0) (1000 255) (0 511)) 1)
        (5 45 ((0 0) (1000 255) (0 511)) 0))
    (EXE 0 6)
        (instr1 0 1 370.0 20)
        (instr1 1 1 425.3 20)
        (instr1 2 2 466.2 20)
        (instr1 4 2 622.3 20)
    (STP)
    (FIM)
```

In this case, the performer is not a musician but a computer which interprets the score and synthesises the music. Readers are invited to refer to my previous book *Computer Sound Synthesis for the Electronic Musician* (Miranda, 1998) for more details about the SOM-A language and sound synthesis in general. The Nyquist language on the accompanying CD-ROM is an ideal tool for composers wishing to work at the microscopic level.

Conversely, at the *building-block level*, composers work with larger musical units lasting several seconds, such as rhythmic patterns, melodic themes and sampled sound sequences. This trend started to gain popularity with the appearance of electro-acoustic music in the 1950s, where pieces were composed by literally cutting, pasting and mixing sections of sounds recorded on tape. This procedure is widely practised today, mostly by DJs and pop musicians working with hip-hop, ambient, jungle and techno styles. Digital sampling and MIDI systems, as well as recorded sound libraries and MIDI files, are widely available to anybody wishing to compose with pre-fabricated sections of music. The Music Sketcher program on the accompanying CD-ROM is a great tool for composing at this level. Music Sketcher provides a library of short musical sections that can be combined in order to form a musical piece. Here the composer can also program the system to apply transformations to these musical sections. The Koan system (also featured on the CD-ROM) is another program for composition with pre-fabricated sections, but here the composer can also create these sections. This book will focus mainly on composition at the level that lies between these two extremes: the *note level*. The reader should bear in mind, however, that most concepts and techniques discussed in the following chapters can be equally applied to other levels.

At the note level, the atomic element of music is a single sound event described by a number of attributes: the *musical note*. Music traditionally has a well defined boundary of abstraction for characterising the musical note in terms of attributes. The musical note has four main attributes: pitch, duration, dynamics and timbre (that is, the timbre of the instrument that plays the note). Tempo, which adds depth and further expression to the notes, is a complementary attribute; it does not make much sense as a single note attribute, but rather as a higher level structure containing various notes (e.g., motif, phrase, track, etc.). Composers are encouraged to think of a musical piece as structures of notes with careful control of their attributes. Here, a composition is often written in a score by means of symbols representing note arrangements, and performers interpret the score by relating these arrangements to gestures on musical instruments. In other words, the symbolic representation of the score does not really present the music, but rather instructions for musicians who learn which actions to perform from these symbols in order to play the music. The inner acoustic features of a note (for example, amplitude of partials and spectral configuration) are not directly relevant for representation on the score. Indeed, at the note level there are no obvious symbols to represent, for example, the spectrum of individual notes; standard acoustic musical instruments were not designed with keys to explicitly change the harmonic spectrum of their sounds. In other words, at the note level composers and performers work with a conceptual model of individual instruments whose boundary of abstraction gives very little room for significant manipulation of the inner workings of the timbre of the notes. It is given by default that different timbres can be obtained through the combination of different instruments playing together. In a way, this is no bad thing; the only drawback is that before the appearance of the synthesiser, only those composers who had access to large orchestras could fully explore the potential of timbre in their pieces. Most programs on the accompanying CD-ROM are designed to work at the note level: Tangent, Texture, CAMUS, Harmony Seeker, M and so forth.

To summarise, the microscopic level deals directly with physical sound attributes, such as frequency, amplitude, spectrum, and so on, which are mainly used for synthesis and sound processing. At the note level, we bundle certain sound attributes together and think of them as a note. Aggregates such as phrases, motives and structures constitute the building-blocks for composition at a higher level of abstraction.

1.2 Time-domain hierarchies

The notion of time-domain hierarchies is complementary to the notion of abstraction boundaries introduced above. Composer Richard Orton proposed four distinct but interconnected musical domains, hierarchically organised with respect to our perception of time: *timbre, frequency, pulse* and *form* (Orton, 1990).

1.2.1 The domain of immediate perception: timbre

The domain of immediate perception is related to musical timbre. Timbre has the finest resolution of the four hierarchical time domains: it is a complex entity formed by a number of components (or *partials*, in acoustics jargon) that are characterised by frequency relationships and relative amplitudes.

Orton suggests that the first aspect of a sound to which we respond is its timbre. This is a controversial statement, but it does make sense to a certain extent: we do seem to react to timbre even before we have the chance to fully process the sounds we hear. This is probably a legacy of a natural survival mechanism that emerged during the early stages of our evolution: as sounds can indicate threat or danger, it is vital to be able to identify the source quickly and to react to what is causing it. In addition to prompt reaction, we also have the ability to track the changes of the spectrum of the sound as it unfolds; this ability was fundamental for the evolution of our linguistic as well as our musical capacity.

Composers working at the microscopic level of abstraction have direct access to the domain of immediate perception. To many contemporary composers, timbre is the ultimate frontier of music that has only recently begun to be explored systematically. This trend is very noticeable in the music of the end of the twentieth century, mostly in France, where composers tended to overemphasise the timbral crafting of the composition (Barrière, 1991). This practice often rendered their pieces overly complex to listen to and appreciate, because the human ear is not as able to make sense of clever sound manipulations at the microscopic level as it is at the note level: our brain seems to rely on timbre more for sound recognition and source localisation tasks than for making sense of structure and form.

1.2.2 The domain of frequency

The notion of *frequency* refers to a sequence of repetitive patterns. In the case of a sound, frequency refers to a sequence of repetitive waveform patterns. Tied to the notion of frequency is the notion of *period*, that is, the time it takes to repeat the pattern. Frequency and period are reciprocally related:

$$Frequency = \frac{1}{Period} \text{ and so } Period = \frac{1}{Frequency}$$

Sound frequencies are measured in cycles per seconds (cps) or Hertz (Hz). On average, the human ear is capable of perceiving frequencies between 20 Hz and 18 000 Hz (or 18 kHz), but this may vary from person to person and with age; the older we get the less we are able to hear frequencies above 8 kHz. Music rarely uses pitches higher than 4000 Hz.

The pitches of musical notes operate in proportional cycles called *octaves*. For example, when the frequency of a note A at

220 Hz is doubled to 440 Hz, one still perceives the same note A, but one octave higher. Conversely, if the frequency is halved to 110 Hz, then the note is perceived one octave lower (Figure 1.1). Our hearing system deals with sound frequencies according to a logarithmic law. Thus, the phenomenon we perceive as pitch interval is defined by a logarithmic process; for example, the distance from 110 Hz to 440 Hz is two octaves, but the frequency ratio actually is quadrupled.

Figure 1.1 The pitches of musical notes operate in proportional cycles called octaves. The phenomenon we perceive as pitch interval is defined by a logarithmic process.

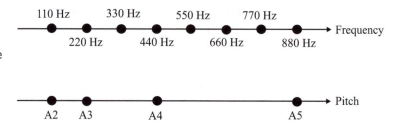

The notion of the octave is said to have been conceived in ancient Greece, where it was discovered that by halving the length of the string of the *monochord* (a one-stringed instrument which was used by the Greeks to explain musical concepts) one could double the octave of the tone. Pythagoras (560–490 BC) is credited for having put forward the notion that a musical scale can be naturally derived within the octave by dividing the string at various points using simple whole-number ratios: 2:1 the octave, 3:2 the perfect fifth, 4:3 the perfect fourth, 5:4 the major third, and so on. These ratios are known as the *Pythagorean ratios of musical consonance* and they are still respected today.

For Pythagoras, numbers and geometry were imbued with intangible mystical connotations. The correlations discovered between whole-number ratios and the musical consonances were seen in terms of a divine revelation of universal harmony. However, in practical terms, the Pythagorean method is not the only method for defining musical scales. Scale systems have emerged and fallen into disuse as musical styles have evolved all over the world. Indeed, non-Western musical scales, for example, operate in completely different ways. The specifics of musical scales are beyond the scope of this book. For a good introduction to the subject, the reader is invited to refer to another book in the Focal Press Music Technology series: *Acoustics and Psychoacoustics* (Howard and Angus, 1996). For the purposes of this book, it suffices to know that one of the most frequently used scale systems today is the *scale of equal temperament* (Figure 1.2), which emerged in Europe in the nineteenth century. In this scale, the octave is divided into twelve equally-spaced semitones.

Figure 1.2 The scale of equal temperament is one of the most popular musical scales of our time.

1.2.3 The domain of pulse

Karlheinz Stockhausen is often referred to as having proposed that both pitch and rhythm can be considered as one continuous time-domain phenomenon. In an article entitled *Four Criteria of Electronic Music*, Stockhausen described a very peculiar technique for producing synthetic tones (Stockhausen, 1991). He recorded individual pulses on tape, from an electronic generator. Then he cut the tape and spliced the parts together so that the pulses could form a particular rhythm. Next he made a tape loop of this rhythm and sped it up until he could hear a tone. Various tones could be produced by varying the speed of the loop. Different rhythms on tape produced different timbres; the components of the rhythmic sequence determined the spectrum according to their individual cycle on the tape loop.

The most interesting aspect of Stockhausen's technique is that it encourages composers to work with rhythm and pitch within a unified time domain. It should be noted, however, that the transition from rhythm to pitch is not perceived precisely. Hence the reason why the domain of pulse is considered here at a higher level in the time-domain hierarchy. Whilst Stockhausen's unified time domain makes perfect sense in quantitative terms, it fails to address the qualitative nature of the human ear: we can hear a distinct rhythm up to approximately 10 cycles per second but a distinct pitch does not emerge until approximately 16 cycles per second. It is not entirely by chance that the categorical differentiation between rhythm and pitch has remained firmly entrenched throughout history.

The domain of pulse therefore lies below the rate of approximately 10 cycles per second but we seem to deal more comfortably with rates that are close or related to the human heartbeat. The extremes of the heartbeat range from approximately 30 beats per minute up to 240 beats. In terms of Hz, a very fast heartbeat would be equivalent to 4 Hz (240 × 1/60). Conversely, a very slow heartbeat would be equivalent to 0.5 Hz (30 × 1/60). Musical rhythms generally fall within this bandwidth; i.e., from four pulses per second (4 Hz) to one pulse every two seconds (0.5 Hz).

The term 'beat' denotes a regular stress or accent that usually occurs in a composition. The notion of *tempo* (a term borrowed from Italian which means time) is associated with the ratio of beat to pulse rate. Until the beginning of the nineteenth century composers indicated this only vaguely on the musical score, using terms such as *adagio* (Italian for slow), *vivace* (Italian for lively), *andante* (Italian for at a walking pace), and so forth. With the invention in the 1810s of Maelzel's metronome, a clock-like device that produces a click according to a specific pre-set rate, the tempo indication became more precise. The scale of measurement is in number of beats per minute and in a musical score this is shown by the abbreviation M.M. (for Maelzel's metronome) and a number; for instance, M.M. 60 indicates a tempo of 60 beats per minute.

Beats can be sub-divided and grouped. The grouping of beats is called *meter* and it is indicated on the score by vertical bar-lines on the staves. At the beginning of the score there is usually a pair of numbers indicating the meter of the piece: the top number indicates the number of beats in a bar and the lower number indicates the reference unit for the beat.

1.2.4 The domain of form

Originally coined by Orton as the *domain of memory*, the *domain of form* involves the placing of musical materials in time.

As with the previous domains of frequency and pulse, musical form also can be thought of as a cyclic succession of events. The first metaphor that comes to mind here is our breath cycle: a succession of inhalations and exhalations that seem to drive the course of our motor and cognitive activities.

The human mind has an incredible ability to impose order on auditory information. Psychologists suggest that we employ mental schemes that guide our auditory system in order to make sense of incoming streams of auditory data such as speech and music (McAdams, 1987). There are controversies as to which of these schemes are genetically hard-wired in our brains and which ones are culturally shaped as we grow up, but we do not need to worry too much about this quarrel here. We will come back to the domain of form when we introduce the notion of cognitive archetypes in the next section.

1.3 Approaching composition

To what extent do composers think differently when composing with computers as opposed to earlier compositional practices, such as the classical picture of the composer working on the piano with a pencil and sheets of music? When engineers use a computer to calculate complex equations or to prototype artefacts, the machine certainly frees their minds to concentrate on the problem at hand, rather than dwelling on the particulars of the equations or drawings. This also applies to musicians, but there are surely other issues to be considered in the case of musical composition, because some composers actually use the computer to perform decision-making tasks. Perhaps more in music than in any other application, the computer can be programmed to be creative and this is a practice that is as old as the computer itself; Hiller and Isaacson's work, already mentioned in the Preface, is a commonly cited example of this practice which dates back to the mid 1950s.

There are basically two different types of software for musical composition: *algorithmic composition* software and *computer-aided composition* software. Whilst algorithmic composition software is programmed to generate music with a certain autonomy, as in the case of Hiller and Isaacson's work, computer-aided composition software serves as a tool to help the composer capture and organise ideas. There is a healthy tension between these two types of software in that most composers interested in composing music with computers tend to find a balance between these two extremes. Nevertheless, this book is more interested in exploring the potential of the former type of software because it tends to embody musical systems and paradigms of a generative nature. MIDI sequencers, software synthesisers, samplers and miscellaneous computer-based studio gear can be cited as instances of the second type of software. Programming environments such as OpenMusic and languages such as Nyquist may also fall into the second category of software for composition.

The following sections introduce three distinct but complementary topics concerning approaches to musical composition using the computer:

1   top-down versus bottom-up,
2   interface modelling, and
3   parametrical thinking.

1.3.1 Top-down versus bottom-up

There are two major approaches to working with a computer as an active partner for composition: the *bottom-up* and the *top-down* approaches. One approach is to engage in improvisation and experimentation with the machine and store promising musical materials. Then at a later stage, these materials are developed into larger passages, musical structures and so forth. This is the bottom-up approach because these smaller sections, created by or with the computer, function as the foundation for building larger musical sections. Higher level musical sections are composed (with or without the aid of the computer) by extending these smaller segments to form the entire piece.

Conversely, composers might prefer a top-down approach, that is, to start by developing an overall compositional plan or computer program beforehand and proceed by refining this plan. This approach forces the composer to be creative within self-imposed formal constraints; for example, the number of sections, the length and character of each, types of generative processes for each section, and so on. The composer may choose to honour the limits or may prefer to change horses in mid-race if these turn out to be inadequate. In this case, the composition process results from the interaction with high-level compositional constraints that are largely operated by the computer.

In most cases, algorithmic composition software tends to support the bottom-up approach whereas computer-aided composition software tends to support the top-down approach. Exceptions to this rule, however, abound here. All the same, as each approach has its pros and cons, composers tend to combine both approaches. Most of the programs on the CD-ROM are designed with the bottom-up approach to composition in mind. Texture, MusiNum, a Music Generator, FractMus and CAMUS are typical examples of this approach. These programs are best used as generators of musical materials to be used in a piece; only those materials that satisfy the composer's aesthetic aims will be chosen. Composers often make further adjustments to these materials in order to fit particular compositional contexts and so on. For example, the author's piece for chamber orchestra *Entre o Absurdo e o Mistério* (Miranda, 2001) was entirely composed using materials generated by CAMUS; two excerpts of this piece are available on the CD-ROM in folder <various>.

For the top-down approach to composition, there is Tangent. Tangent works by creating sequences of musical passages, for each of which the composer can specify different generative attributes. Here the composer is encouraged to think in terms of the relationships between these passages from the point of view of musical form: e.g., 'passage B will have more rhythmic contrast and trills than passage A', 'passage C is longer and more homogeneous than passage D', and so on. Also, OpenMusic features an interesting tool called *maquette*, which functions as a kind of canvas where the composer can place icons representing programs for generating musical passages, samples, MIDI sequences, and so on. Both Tangent and OpenMusic are introduced in Chapter 8 and can be found on the accompanying CD-ROM.

1.3.2 Interface modelling

An interesting approach to composition with computers has recently been proposed by Mikhail Malt (1999). Malt breaks the composition process into two main activities, namely, *conceptual activities* and *writing activities*. These are not necessarily sequentially distinct activities, but rather concurrently interrelated in the sense that a composition results from the interaction of the two.

Conceptual activities involve the specification of prescriptive orientations for the piece. After deciding upon an abstraction boundary, the composer then elaborates the strategies, the rules, the conceptual and metaphorical subtracts of the composition, and so on. The format of these conceptualisations will vary enormously from composer to composer: they can be in the form of sketches, drawings, diagrams, written rules, mathematical formulae, etc. Specific musical styles will support different conceptual specifications; for example, the harmonic constraints for a piece of *tonal music* (Schoenberg, 1967), the sequence of operations for a piece of *serial music* (Boulez, 1963) or the probability formulae for a piece of *stochastic music* (Xenakis, 1971), to cite but a few. Other less formal issues may also be at stake here, such as aesthetic preferences or the composer's political preferences (Cardew, 1974), for example.

At the writing stage, the composer then materialises his conceptualisations into music. Writing music normally involves a great deal of decision making, ranging from the choice of notes and keys, to the choice of instrumentation. Here, writing music is an act of expressing or representing concepts and musical ideas, not necessarily in the graphical sense. Indeed, according to Malt, there should be no distinction between writing a piece on paper using a pencil and working directly with sound recordings on a computer.

The dichotomy between the conceptual versus writing activities naturally leads to the notion of *compositional models*. In this context a compositional model is a mediator between the abstract and the real: it is a means whereby composers can turn their concepts into music (Figure 1.3). Since computers are excellent at modelling tasks, it comes as no surprise that the computer can be effectively employed as a tool to turn ideas into music.

Figure 1.3 A composition model mediates abstract ideas and their musical realisation.

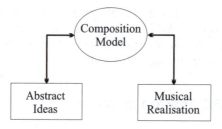

Malt identifies four categories of compositional models: *logical, analogical, metaphorical* and *aesthetical*. It is beyond the scope of this book to introduce a thorough discussion on these four models; suffice to say that the computer works at its best for the first two categories. Computers are better employed for those musical models involving scientifically or pseudo-scientifically inspired musical thought; hence this book's tendency to support logical and analogical models.

1.3.3 Parametrical thinking

The emergence of parametrical thinking in music has been of paramount importance for the development of computer-based composition systems. Parametrical thinking in general is a natural consequence of the increasing trend towards computer-oriented systematisations of our time and it is highly inspired by *Cybernetics* (Wiener, 1948). In *Cybernetics*, a parameter is one of the variables that controls the outcome of a system. In our case, the output is music and the parameters refer to controllable musical attributes whose values can vary within upper and lower limits. As an example of a cybernetic musical system, imagine a musical box furnished with a number of buttons that can be turned from one extreme of the dial to another; each button controls a specific attribute of the composition (pitches, tempo, etc.) and these buttons can be combined to form meta-buttons, and so on.

The parametrical approach to composition has led composers to think in terms of parametrical boundaries and groupings that probably would not have been conceived otherwise. Classic examples of parametrical thinking can be found in the book *Formalized Music*, by Iannis Xenakis (1971), particularly where he describes the algebraic method he devised for composing *Herma*, for piano (Xenakis, 1967). An example of a Xenakis-like compositional method will be given in Chapter 2.

It may be difficult for a computer musician to imagine how composers of previous centuries could operate without thinking parametrically, but in a sense we could say that parametrical thinking goes back to medieval music where rhythms and note sequences were created separately and then combined at a later stage. For example, pitches could be assigned to a text, forming what was referred to as *color*, and then the notes were made to fit a rhythmic pattern, referred to as *telea*. If the color and telea sequences were of different lengths, then they were rotated in combination against each other many times until their ends matched. Perhaps the main difference between cybernetic parametrical thinking and medieval practices is that the former considers the attributes holistically.

The range of attributes available in a musical system is an important aspect to consider in a composition because it contributes to the type of unfolding one wishes to imprint onto the piece. The smaller the inventory of all possible intervals, chords, rhythms, and so on, the fewer the combinatorial possibilities. The fewer the combinatorial possibilities, the easier it is for a composer to handle the material, but the outcome will tend to be either short or repetitive (an example of combinatorial composition is discussed in Chapter 7). The range of possibilities can of course be augmented by increasing the inventory, but the larger the inventory the greater the chances of producing pieces beyond the human threshold of comprehension and enjoyment. For instance, it is perfectly within our hearing capacity to use scales of 24 notes within an octave but more than this would only create difficulties for both the composer and listener.

Regardless of musical style, an increase of focus on certain parameters in a musical passage should be accompanied by a decrease of focus on others. The human mind has an attention threshold beyond which music is most likely to be perceived as lacking order or direction. For instance, if the intention is to focus on melodic lines, then one should refrain from making too many contrasting timbre variations while the melodies are unfolding, otherwise the timbre may distract attention away from the pitch. However, melody alone would not normally keep the listener's attention for long because a piece of music needs some variety and contrasting elements from time to time. One of the most difficult issues in composition is to find the right balance between repetition and diversity: a well-balanced piece of music should unfold naturally to listeners and this is very difficult to achieve. In most Western musical styles, this unfolding results from a network of relationships between musical attributes that combine in different ways to create situations where contrasts can take place. Such contrasts are normally preceded by a preparatory stage, which builds upon the listener's expectations.

13

The main problem with the parametrical approach to music is that the correlation between the parameters, their values and the musical effect heard is not always obvious. Moreover, the role of a single parameter is often dependent on many different musical attributes that have a subtle, and occasionally unknown, interrelationship. For instance, imagine that our musical box allows us to speed up a melody simply by tweaking a tempo button: this would imply changes not only to the triggering time of the notes, but also their duration, articulation, phrasing and so forth.

1.4 Cognitive archetypes

1.4.1 Metaphorical associations

Our cognitive capacities do not work in isolation from one another. Whilst our ability to infer the distance of a sound source is tied to our notion of timing and space, our tendency to associate colours with temperatures (e.g. red and blue with high and low temperatures respectively) seems to be tied to our notions of fire and ice, to cite but two examples. In this context, music may be regarded as the art of invoking cognitive phenomena at one of the most deep and sophisticated levels of human intelligence: the level of *metaphorical associations*.

One of the most important metaphors evoked by music is the notion of *motion*. We suggested earlier that a musical piece is the result of the dynamical unfolding of various musical attributes at different levels, namely levels of timbre, frequency, pulse and form. Here we build upon this notion by conjecturing that our ears naturally take in this dynamical unfolding as sequences of sonic events that seem to move from one lapse of time to another. When we engage in more sophisticated listening experiences we probably employ a number of cognitive strategies that are similar to those we employ to understand the events that take place during the course of our lives.

Comparable strategies also seem to be employed when we read a text. Although of a significantly different nature, a piece of music and a novel, for example, are both multi-layered systems in which sequences of signs are used to convey emotions, ideas, and so on, according to specific conventions; e.g. harmony and counterpoint rules in music, and grammars and lexicon in language. Hence, it comes as no surprise that music and literature appear to share similar underlying structural principles (Jordan and Kafalenos, 1994), at least in the Western tradition.

Due to the fact that music does not have the extra textual refer-
ents of literature, the musical materials themselves must
somehow enable listeners to infer syntagmatic and associative
values to groups of sonic events at various levels. Music is not
just any random sound arrangement; sound arrangements need
something else to sound musical. After the work of composers
such as John Cage and the like (Revill, 1992), this argument may
appear to be adding insult to injury, but let us adopt a more
pragmatic attitude here: the ultimate challenge for composers
is to identify what this *something else* is that will draw the
attention of the listeners at specific moments. If this boils
down to randomness and non sense noises, then so be it.
Notwithstanding, most composers would agree that musically
meaningful sound events, or *musical syntagms*, should convey
cues that enable listeners to infer organisational structures from
them. These cues are normally associated with the style of the
piece: a listener who is familiar with the style relies upon
associative evaluations to group musical syntagms according to
listening strategies that have proven successful for other pieces.
The criteria by which people normally group these syntagms are
based upon tonality, texture, timbre and rhythmic pattern, and
these often occur in combination.

A musical style thus defines a system of syntagmatic relations
in musical material that helps listeners to infer their musical
structures. The grouping of smaller musical units is the first
stage of the process that leads the listener to mentally build
components of large-scale musical forms. Both narrative texts
and musical compositions are processes that are presented in
small discernible units, strung together in a way that leads the
listener to mentally build larger sub-processes at the syntagmatic
level, resulting eventually in the creation of a complex structure
in time. In this sense, one could think of a composition as
musical structures that convey some sort of narrative whose plot
basically involves transitions from one state of equilibrium to
another. Equilibrium here is meant in the cybernetic sense
(Wiener, 1948): it describes a stable but not static relation
between musical elements. In music, two states of equilibrium
are often separated by a stage period of imbalance (Figure 1.4).

Composers wishing to develop complex musical forms and
large-scale pieces need to devise strategies in order to provide
reference points for the listener to hold on to the piece, otherwise it may lose its sense of unity. Different styles and different
composers have developed their own methods. In the Baroque
period, for example, identical repetition of passages in ternary
form A-B-A was very common (Figure 1.5). Perhaps one of the

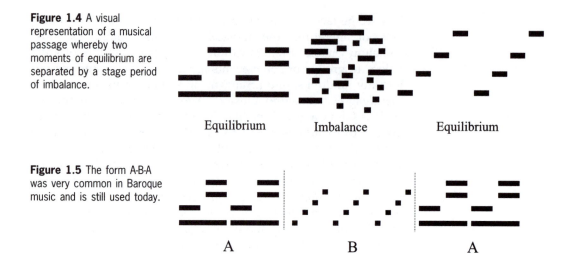

Figure 1.4 A visual representation of a musical passage whereby two moments of equilibrium are separated by a stage period of imbalance.

Equilibrium Imbalance Equilibrium

Figure 1.5 The form A-B-A was very common in Baroque music and is still used today.

A B A

most sophisticated forms of classical music is the *sonata form*, as perfected by the eighteenth century composer Joseph Haydn and still used today (see Figure 3.12 in Chapter 3).

The importance of structure, syntagmatic relations and associative values in music is highly debatable. With regard to these notions, there is an extreme school of thought in musicology purporting that in order to understand and appreciate music one must be able to naturally parse it and understand its structure: notes, phrases, motives, voice and so on. Readers who are interested in studying these themes further are invited to refer to the work of Meyer (1956), and Lerdhal and Jackendoff (1983).

1.4.2 Elementary schemes

Paul Larivaille (Larivaille, 1974) proposed a five-stage elementary scheme for narrative in literature that is a useful starting point to study some standard musical forms. The five stages are:

1   initial equilibrium
2   disturbance
3   reaction
4   consequence
5   final equilibrium.

We can see this scheme in Homer's epic *The Odyssey*, in which the author recounts the adventures of Odysseus on his voyage from Ogygia to Ithaca. In a tonal musical context, for example, the *initial equilibrium* would correspond to the initial section at the tonic key. The arrival of elements that conflict with the tonic key, such as a dissonant note, would then introduce a *disturbance* in the initial state of equilibrium. Disturbance is invariably followed up by a *reaction*, such as modulation to the dominant key, for example. The arrival at the dominant key is the *consequence* of the action and the settlement at this key establishes the *final equilibrium*. However, pieces of music would seldom end here. The final equilibrium could indeed be disturbed again, thus recommencing the whole cycle, and so forth. In works of greater structural complexity, these elementary sequences can be combined to form various layers, which could follow the same type of narrative form. For instance, a sonata would normally have an exposition of a theme in the tonic key (an initial *meta-equilibrium*) followed by a dominant section (the *meta-distrubance*), the development section of the sonata (a *meta-reaction*) and the recapitulation section (the *meta-consequence*). The classical sonata regularly ends with a coda (the *final meta-equilibrium*). This process of building higher-level musical structures with the same form of its embedded components resembles the iterative processes behind fractal structures; fractal music will be discussed in Chapter 4.

Note that interesting forms can be created by extrapolating the sequential nature of Larivaille's stages. One could create *embedding forms*, by inserting a second sub-sequence during a pause in the main sequence, and/or *alternating schemes*, by maintaining more than one sequence simultaneously with or without interruptions, to cite but two examples. What is important to bear in mind here is that the succession of sound events in a piece should give the listener a sense of direction.

As mentioned earlier, music psychologists suggest that our sense of musical direction may be prescribed by schemes of auditory expectation that we employ to predict which sound event will occur next in a musical sequence. These schemes embody some sort of predetermined grouping that joins the elements into well-formed units in the listener's brain. These musical schemes are probably culturally shaped phenomena that emerge from the various kinds of associations that people make between music and extra-musical contexts, such as the associations between colour and temperature mentioned earlier. There are however some schemes that are less culturally dependent than others, which can be found in the music of almost every culture on earth (Reck, 1997). The most notorious examples of these are the *responsorial expectation*, the *convex curve* and *the principle of intensification* schemes.

One of the most typical musical schemes we seem to employ in musical listening is the responsorial auditory expectation scheme whereby people normally seek to identify musical passages in which sound events seem to call and answer one another. Most interestingly, this phenomenon often manifests itself in a convex arch-like shape, which possibly mirrors the rising and falling patterns of breath.

The breath metaphor also applies to the relationship between various simultaneous voices and their harmonic organisations. In Renaissance music, for example, polyphonic melodies have their own shape and rhythm, but the individual voices contribute to the overall harmonic convex structure. This structure normally follows Larivaille's stages previously mentioned (Figure 1.6). This interplay between melodic phrases and harmonic structure that began to appear in Renaissance music fostered a number of composition techniques throughout musical history; for example: time-displacement of musical phrases (an effect known as *musical canon*) and note progression in different proportions (two voices start together but the time displacement occurs gradually).

Figure 1.6 The overall arch-like shape of the breath metaphor normally follows the five-stage elementary scheme proposed by Larivaille.

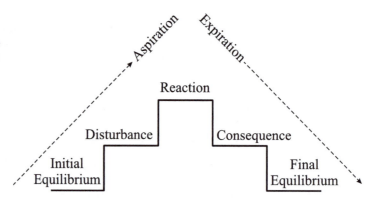

The arch-like convex curve scheme provides an intuitive guide by which one can predict the evolution of given parameters in a process: there is normally one stage where the music ascends towards a climax, followed by a descending stage (Figure 1.7). Both the ascending and descending stages should normally evolve smoothly and at gradual intervals. Smaller intervals here often succeed large ones during the ascending phase and the reverse occurs during the descent. This scheme may apply to all domains of music, ranging from the domain of timbre to the domain of form. Historically, however, it has been consciously used by composers in the domain of form: one of the first composition techniques that aspiring composers learn at the Conservatory is to construct musical phrases in the style of Palestrina by starting with slow notes at the low-pitch register, moving towards faster notes at higher-pitches and then conclude with slow and low-pitched notes.

Figure 1.7 The convex arch-like scheme is one of the strategies that people employ for predicting the fate of musical sequences.

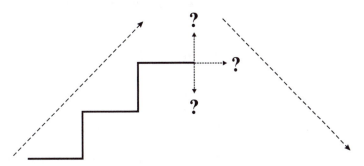

Convexity is a means to achieve well-balanced musical passages and can be used to forge those moments of equilibrium mentioned earlier. The opposite of convexity should thus produce moments of tension: linearity (i.e., no changes), sudden changes, abrupt zig-zags and concavity are but a few examples that can produce tension in a piece of music (Figure 1.8).

Figure 1.8 Unfulfilled predictions often cause musical tension.

As for the principle of intensification, it is commonly employed by composers to create tension and to give a sense of direction for the musical unfolding. The principle of intensification may apply to various time spans and may be expressed in various ways in different parameters through contrasts within a continuous scale ranging from the few to the many, low to high, non-rhythmic to rhythmic, slow to fast, small density to large density and so on (Figure 1.9).

Intensification is normally used to build tension towards a climax, prompting the rise of a convex curve that may or may not be resolved. Some non-Western musical cultures often employ intensification to forge convexity.

Figure 1.9 An example of the principle of intensification.

1.5 Concluding remarks

This chapter has presented a brief introduction to several varied topics of interest concerning compositional approaches and paradigms. The comparison of music and literature in the last sections is a controversial issue that divides musicologists all over the world. It is not the intention of this book to take sides; rather, these theories can be useful for devising compositional systems despite their musicological contention.

By way of conclusion, one of the main messages that this chapter conveys is that contemporary musical composition can be approached from many different levels and angles. For didactic reasons this book will focus most of its discussion on the note level and from an often traditional angle. Notwithstanding, the reader should always bear in mind that the techniques and concepts presented in the forthcoming chapters may, in most cases, be equally applied at the microscopic and at the building-block levels of composition. The computer can process musical material, be it samples or highly abstract representations, at different levels: from sound synthesis to musical form.

One of the main caveats of computer music is that musicians tend to concentrate on the details of specific algorithms, forgetting that neither clever synthesis techniques nor complex musical grammars can guarantee good compositions. But again, a flautist may team up with a percussionist because neither can play both instruments at once: it is with this spirit that composers should team up with computers.

Leonard Bernstein's book *The Unanswered Question* (Berstein, 1976) is suggested as an interesting complement to this chapter.

2 Preparing the ground

The majority of 'computer music' today is undertaken using commercial software that does not require knowledge of mathematics, algorithms, etc. However, such systems have a limited scope: sequencers, accompaniment systems, recording and playback gear. This book intends to broaden the horizons for those musicians who want to go further: hence the need to become familiar with fundamental computer music concepts and be prepared to use systems that either embody some interesting compositional model, allow for programming, or both.

Whilst geometry, numerical proportions and esoteric numbers served to scaffold great compositional minds in past centuries (James 1993), logic, probabilities, set theory and algorithms inspire some contemporary composers. This chapter begins by introducing fundamental mathematical concepts that are crucial for mastering computer music, from discrete mathematics and set theory, to logic and formal grammars. Then, it gives a primer in general computer programming with examples followed by an overview of two historical landmarks for the development of computer music: Schoenberg's serialism and Xenakis' formalised music.

2.1 Elementary discrete mathematics

The term discrete is defined in the Oxford English Dictionary as 'separate, detached from others, individually distinct; opposed to continuous'. Discrete mathematics is therefore appropriate when objects are counted rather than weighed or measured and it often involves relations between sets of objects. Discrete mathematics is the essence of computation because the objects a computer manipulates are discrete, e.g. on/off or 0/1 values in memory.

Mathematical modelling is perhaps one of the most powerful scientific and designing tools created by mankind. The fact that the laws of physics can be expressed in mathematical terms enables scientists, engineers and designers to build simulations and artefacts that would otherwise be impossible to conceive. A vast number of mathematical modelling techniques exists to assist the design of aeroplanes, vessels, cars, bridges, and so on. The choice of a suitable technique is crucial for obtaining the right assistance. Two major mathematical modelling paradigms of musical interest are introduced below: *algebraic modelling* and *graphs-based modelling*.

2.1.1 Algebraic modelling

In order to illustrate how an algebraic model can be built, let us define a simple model for the analysis of the prices of compact discs (CDs) at a certain record shop. Imagine that one person, represented here as P_1, bought one CD of classical music and two CDs of dance music and paid £17. Then, another person, P_2, bought three CDs of classical music and one CD of dance music and paid £21. The objective of this exercise is to deduce the individual prices for each CD of classical and dance music. If we establish that the price of one classical CD is denoted by x and the price of one dance CD is denoted by y, then we can make two mathematical statements as follows:

P_1: $1x + 2y = 17$

P_2: $3x + 1y = 21$

These two statements constitute the kernel for an algebraic model where the CDs are represented in terms of the total sum paid by the customers. Notice that many assumptions have been made here; for example, it is assumed that all classical music CDs are the same price and so are the dance music CDs. Also, in the modelling process we did not consider the demographic details of the customers; it is assumed that this information is not important for the model.

Let us now examine how the individual CD prices can be derived from the model. In essence, algebra works by juggling with the terms of the equations with the objective to deduce values for their variables. For instance, by multiplying the second equation by two we obtain the equation:

$$6x + 2y = 42$$

Next, by subtracting the first equation from the third equation, we obtain a value for x. This value corresponds to the price of a single classical music CD, as follows:

$$5x = 25$$

$$x = 25/5$$

$$x = 5$$

Next, the value of y, that is the price of a dance music CD, can be obtained by substituting x in the equation above:

$$5 + 2y = 17$$

$$2y = 17 - 5$$

$$2y = 12$$

$$y = 12/2$$

$$y = 6$$

From the results of these operations we can finally conclude that the price of one CD of classical music is £5 and the price of one CD of dance music is £6. This model is discrete in the sense that it deals with distinct entities (CDs, people, prices) and that the solution for the problem can be found after a number of finite algebraic steps. The model is extremely simple, but it nevertheless has some predictive power for calculating the bill for people who buy classical and dance music CDs in this particular shop. What is important to consider here is that the mathematical operations were performed on an abstract model of the real facts. This is where the power of mathematical modelling for music lies.

2.1.2 Graph-based modelling

Let us now consider the following example: imagine a list of geographical places and that each pair of places is linked by a route; the distances of each route is known. A table containing the distances in metres is given as follows:

Place A				
2810	**Place B**			
2280	1670	**Place C**		
3780	1875	3800	**Place D**	
3940	2680	2610	3080	**Place E**

The task here is to find a set of routes that satisfy the following properties:

(a) the routes should connect all places (directly or indirectly)
(b) the total distance of the routes should be the shortest possible

Computers are excellent for dealing with this kind of problem thanks to graphs-based representation techniques. The graph representation of our problem is shown in Figure 2.1. In graphs jargon, the places are represented by *vertices* and the routes are represented by *edges* (or *arches*). Each edge has a *weight*, which in this case corresponds to the distance of the respective route.

Figure 2.1 An example of a graph.

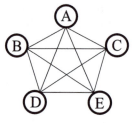

The graph representation facilitates the definition of rules for deriving information that satisfy some given conditions; in this case, the set of edges with the minimum distance. For example:

Step 1: start with an empty set
Step 2: select the edge with the lowest weight that has not yet been selected
Step 3: however, an edge is not eligible if it could form a closed path with the other edges in the set
Step 4: if the edges in the set still do not link all vertices of the graph, then go back to step 2; otherwise the current set of edges contains the routes, as required

In order to better assess the outcome from the application of these rules, the set could be represented in the form of a *tree*; another form of representation that is frequently used in computer science. Note in Figure 2.2 how the edges have been arranged in ascending order of distance values. This might not be the best algorithm, but it is certainly one of the simplest.

Figure 2.2 The solution for a graph-based problem can be represented as a tree.

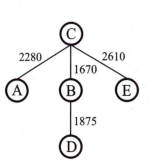

2.2 Fundamentals of set theory

One of the most important notions of discrete mathematics is the notion of a set: a set is a collection of items, technically called the *elements* of the set. The strength of set theory is that it enables one to work with various branches of mathematical modelling under one umbrella, thus facilitating the connection between different domains, such as for example, computing and music.

Set theory is an ideal starting point for tracking problems which involve large amounts of data that need to be examined from different structural viewpoints. A typical use of set theory is for computer applications that involve large databases. Here, set theory provides efficient ways for describing the various categories of information by retrieving and combining clusters of data. For example, a single file of a database may contain the set of all pieces of music composed by a certain composer; another file may contain a collection of musical instruments, and so on. This database could be programmed to establish the relationship between the files. For instance, a relationship labelled as 'employed in' could be established between members of the set of musical instruments and members of the set of pieces of music (Figure 2.3). Many other relationships could be specified; another example could be an orchestration rule defining which elements of the instruments set combine well in specific musical circumstances.

Figure 2.3 Set theory provides an efficient way of structuring data.

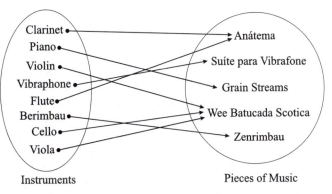

Instruments   Pieces of Music

One important condition of set theory is that the elements must be unambiguously assigned; that is, they must differ within the category they represent. For example, the set of all customers of a bank cannot contain repeated account numbers. Although there is a branch of mathematics dedicated to fuzzy situations, where some items are neither definitely in nor definitely out of the set, the all-or-nothing assignment is adequate for the majority of set applications.

As far as mathematical notation is concerned, sets are conventionally denoted by capital letters and the elements of a set are denoted by lower case. The symbol $\in$ means 'belongs to' or 'is an element of' and the symbol $\notin$ means 'does not belong to' or 'is not an element of'. For example, $x \in A$ reads as 'the element x belongs to the set A', whereas $y \notin B$ reads 'the element y is not an element of set B'.

The elements of a set may be listed individually, separated by commas and surrounded by brackets; e.g. {Antunes, Arcela, Carvalho, Garcia, Iazzetta, Lintz-Maues, Richter, Silva} to denote a set of Brazilian composers. However, this method is not adequate for large sets. If the list of elements has an obvious pattern, then we can omit some of the elements. For instance, a set consisting of all multiples of three up to 30 could be written as follows: {3, 6, 9, ..., 30}. In most cases, however, large sets are better represented by stating the properties of its elements: {all multiples of 3 between 1 and 30}. Mathematically, this is written more elegantly as: $S = \{x \mid x = 3n, n$ an integer and $1 < x \leq 30\}$. In plain English, this statement reads as follows: 'S is the set of elements x, such that x is equal to three times a certain integer n, and x should be greater than one and lower than or equal to 30'. Clearly, the mathematical form is far more efficient for expressing this sort of thing.

Unless otherwise stated, the order in which the elements of a set occurs does not matter. Thus, {12, 24, 48} is identical to {24, 48, 12}. There are cases in which the order is important to consider; these cases will be introduced later.

In set theory, all elements of a set should normally belong to a domain set technically referred to as the *universal set*, represented as U. The sets of all integer numbers and the set of all real numbers are two examples of a universal set. Note, however, that in practice there is no such thing as a universal set; its definition will depend on the particular application of the set theory. For example, the set of all musical notes could be the U in a musical composition system. Another important set to consider here is the empty set; that is a set that contains no elements. The empty set can be denoted either by { } or by $\emptyset$.

A set A is regarded as a subset of a set B if every element of A is a member of the set B. This is represented as follows: $A \subset B$. Conversely $A \not\subset B$ denotes the case where A is not a subset of B. Thus, {1, 3, 5} $\subset$ {1, 2, 3, 4, 5} but {2, 4, 6} $\not\subset$ {1, 2, 3, 4, 5}. From all definitions we have seen so far, it follows that: for all sets A, $A \subset U$ and $\emptyset \subset U$. Also, set A is equal to set B if $A \subset B$ and $B \subset A$.

Finally, the set of all subsets of a set is technically called the *power set*. For example, the power set for $S = \{1, 3, 5\}$ will have the following elements: $\varnothing$, $\{1\}$, $\{3\}$, $\{5\}$, $\{1, 3\}$, $\{1, 5\}$, $\{3, 5\}$ and $\{1, 3, 5\}$; note that sets can contain other sets as elements.

2.2.1 Set operations

Set operations are often represented with the aid of Venn diagrams. These diagrams are useful to quickly visualise relations between sets or to get an overall idea on how to tackle a particular problem. Bear in mind, however, that Venn diagrams are not suitable for handling complex set operations. Figure 2.4 illustrates the Venn diagram representation of the relation $A \subset B$.

Figure 2.4 The Venn diagram representation for A B.

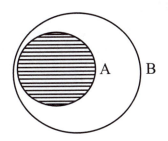

Basically there are four set operations for combining the elements of sets into new sets: *union, intersection, difference* and *complement*.

The union of two sets A and B is written as $A \cup B$ and it consists of the set of all elements of both A and B (Figure 2.5). For example if $A = \{1, 3, 5\}$ and $B = \{2, 4, 6\}$, then $A \cup B = \{1, 2, 3, 4, 5, 6\}$.

Figure 2.5 The union of two sets.

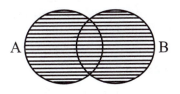

The intersection of two sets A and B, written as $A \cap B$, is the set containing all elements that A and B have in common (Figure 2.6). For example, if $A = \{2, 4, 6\}$ and $B = \{3, 6, 9\}$, then $A \cap B = \{6\}$. If both A and B have no elements in common, then $A \cap B = \varnothing$.

Figure 2.6 The intersection
of two sets.

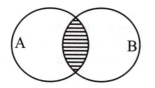

As for the difference of two sets, denoted either as $A - B$ or as $A \setminus B$, it consists of all the elements of A which are not in B (Figure 2.7). For example, if $A = \{2, 4, 6\}$ and $B = \{3, 6, 9\}$, then $A - B = \{2, 4\}$.

Figure 2.7 The difference of
two sets.

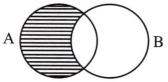

Finally, the complement of a set A is denoted as A' and it consists of all the elements of the respective universal set which are not contained in A (Figure 2.8); that is $A' = U - A$. For example, if we define $U = \{1, 2, 3, 4, 5, 6, 8, 9, 10\}$ and $A = \{1, 3, 6\}$, then $A' = \{2, 4, 5, 7, 8, 9, 10\}$.

Figure 2.8 The complement
of a set.

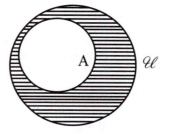

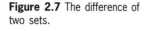

The OpenMusic package on the accompanying CD-ROM (refer to Chapter 8) provides various programming tools for implementing set operations on musical data. These are well documented in the reference manual.

2.2.2 Set algebra

The basic set operations introduced in the previous section take only two operands, or sometimes even one. In order to understand how these four basic operations can be applied to solve more complex set equations, it is necessary to revisit some facts about arithmetic which we generally take for granted.

Addition and multiplication are *commutative* in the sense that the order in which the numbers are added or multiplied does not make any difference to the result; e.g. $2 \times 3 = 3 \times 2$ and $2 + 5 = 5 + 2$. Set unions and intersections are similarly commutative:

$A \cup B = B \cup A$
$A \cap B = B \cap A$

Another important fact is that multiplication is *distributive* over addition in the sense that $2 \times (3 + 5) = (2 \times 3) + (2 \times 5)$. In set theory, unions and intersections are distributed over one another. For example:

$A \cup (B \cap C) = (A \cup B) \cap (A \cup C)$
$A \cap (B \cup C) = (A \cap B) \cup (A \cap C)$

Also, there is the *associative* law of addition and multiplication which says that $(2 + 3) + 5$ is equivalent to $2 + (3 + 5)$ and so is $(3 \times 4) \times 2$ equivalent to $3 \times (4 \times 2)$. This property also applies to the unions and intersections of set theory:

$A \cup (B \cup C) = (A \cup B) \cup C$
$A \cap (B \cap C) = (A \cap B) \cap C$

In addition to the commutative, distributive and associative properties there are six others to consider, as described in Table 2.1:

Table 2.1

Complement	$A \cup A' = U$	$A \cap A' = \varnothing$
Idempotency	$A \cup A = A$	$A \cap A = A$
Identity	$A \cup \varnothing = A$	$A \cap U = A$
Zero law	$A \cup U = U$	$A \cap \varnothing = \varnothing$
De Morgan's law	$(A \cup B)' = A' \cap B'$	$(A \cap B)' = A' \cap B'$
Involution	$(A')' = A$	

2.2.3 Selection and combination

An important aspect of set theory concerns methods for selecting and/or combining set elements into subsets.

Let us imagine the case of a large composition project divided among a team of eight composers over the Internet. The dynamics of the work is as follows: there is no main project leader, but each composer plays the role of creator and double-checker. Once each section of the work has been completed by a creator, it will be passed to a different composer – the double-checker – for proofing and amending. The question is how many combinations of creators and double-checkers are possible for any one section of the piece? If one decided to make the selection by pulling the names from a box and considered that the first one was the composer and the second the double-checker, then the answer would be 56. In this case the order in which the composers are selected is important because when composer X is the creator and Y is the double-checker, it is clearly not the same case as when Y is the composer and X is the double-checker.

There are cases, however, where the order in which the elements of a group are selected is not important. For example, if the objective were to estimate how many pairs of composers could be formed, then the answer would be 28; the pair AB and BA would be considered equivalent here. These two simple examples well illustrate the fundamental difference between unordered and ordered sets; sometimes ordered sets are referred to as *n-tuples*. Generally, if the n-tuple of items $[a_1, a_2, a_3, ...]$ constitutes a different solution to a problem from items $[a_2, a_1, a_3, ...]$, then the order of the elements does matter.

Another important factor to consider when selecting elements from a set is whether or not repeated selections of the same item are allowed. Of course, in the case of the composition where there must be two distinct people, it is evident that the same composer cannot be paired to him/herself. But there are situations where this is possible. To summarise, problems involving the combination of the elements of a set can thus be classified into four categories:

1   Order does not matter and items can be repeated
2   Order does not matter but items cannot be repeated
3   Order matters and items can be repeated
4   Order matters and items cannot be repeated

A general terminology for referring to selection problems is defined as follows: n represents the amount of items from which the selections will be made and r the number of items in the group of selected items; r does not necessarily need to be smaller than n. Back to the composition project, selecting two people from a group of eight gives eight options for selecting the first composer, but gives only seven options for selecting the second because one has already been chosen. This gives a total of 56 combination, that is, 8×7. If it was necessary to choose three at each time, then the options would be firstly eight, then seven and then six: $8 \times 7 \times 6$. Hence, for the case where r items are to be selected from n, the result will be $n \times (n - 1) \times (n - 2) \times ...$ and so on. Because there are r factors in the expression this is normally referred to as a *factorial* and it is written as $n!$; for example $3! = 3 \times 2 \times 1$.

Mathematicians use the term *permutation* for the kind of selection where order matters and repeats are not allowed and the formula for calculating the number of permutations in a given set is as follows:

$$^np_r = \frac{n!}{(n-r)!}$$

This formula states that the number of possibilities for selecting r items from n, where the order matters and repeats are not allowed, can be estimated by dividing the factorial of n by the factorial of $n - r$; e.g. if $n = 8$ and $r = 2$, then $8!/6! = 40320/720 = 56$.

Conversely, the case of selection where order does not matter and repeats are not allowed is referred to as *combination* and the formula is as follows:

$$^nC_r = \frac{^np_r}{r!}$$

This formula states that in order to calculate the number of combinations one simply divides the number of permutations by the factorial of the number of items per selection. Back to our example, the number of combinatorial pairs of composers is achieved by dividing 56 by 2!, which amounts to 28. Hence, *AB* and *BA* are regarded as equivalent in combination but different in permutation, and in the case of pairs, there are only half as many non-ordered possibilities as there are permutations.

As for the cases where the order matters but there can be repetitions, the formula is extremely simple: n^r. However, the remaining case where the order does not matter and there can be repetitions, the complexity of the formula increases slightly:

$$\frac{(n+r-1)!}{(n-1)! \times r!}$$

The OpenMusic package provides various programming tools for implementing combinatorial processes for musical composition. Please refer to Chapter 8 and to OpenMusic's reference manual available on the CD-ROM for more details.

2.3 Basics of logic

This section introduces two fundamental concepts of mathematical logic: *Boolean algebra* (after George Boole, one of the founders of mathematical logic) and *logical deduction*.

The building blocks of logic are *statements*. Statements are simple declarative sentences which are either true or false, but cannot be both at once. This is analogous to set theory where an element either belongs to a set or not. As with the case of set theory, there is a growing interest in finding ways to deal with fuzzy logic statements ('may be true', 'could be false', etc.).

Instead of taking numerical values, Boolean quantities take only two values: 1 (for *true*) or 0 (for *false*), also represented as T and F, respectively. For example, the statement 'The earth is flat', would be true if the earth were flat, but in fact this sentence is false. This statement is a constant statement; that is there are no variables in it. However, computer programming normally requires the use of statements which have variable values. As an example suppose that $P(x)$ corresponds to the proposition 'x is an even number'. As it stands, it is not possible to know whether this is true or not. But if one sets $x = 2$, then the statement is true; conversely, by setting $x = 3$, then it is false.

As with the case of set theory, it is often inconvenient to write logical statements literally. In order to facilitate this task, there are a number of conventions for abbreviating them. The logic symbols that are normally used are:

$\forall$ = stands for 'for all' or 'for every'
$\exists$ = stands for 'there exists'
$\neg$ = symbol to indicate negation (e.g., $\neg a$ = not a)
$\wedge$ = means 'and'
$\vee$ = means 'or' or 'possibly both' (also known as 'inclusive or')

Thus, instead of writing statements such as 'all values of x such that x^3 is a non-negative number', one could simply write: $\forall x \mid x^3 \geq 0$. Another example: 'for every number x, there exists a number y such that y is bigger than x' could be expressed as $\forall x, \exists y \mid y > x$.

The last three symbols of the above list are for logical operations. These operations by and large resemble the algebraic operations of set theory; e.g., an element belongs to $A \cup B$ if it is in $A \vee B$; an element belongs to $A \cap B$ if it is in both $A \wedge B$; and so on. Here, compound statements resulting from the use of these logical operators can only value 1 or 0, just as if they were single statements, but the truth or falsehood of a compound statement will depend upon the truth or falsehood of each single statement of the compound. If a = 'Beethoven composed sonatas', b = 'Beethoven was born in Tokyo' and c = 'A sonata is a musical form', then a and c are true but b is false. In logic, the truth of a compound statement involving the operator '∧' requires both components of the expression to be true; hence, $a \wedge c = 1$ but $b \wedge c = 0$.

The rules for deducing the truth or falsehood of compound Boolean expressions are summarised in the form of *truth tables*. As with an algebraic expression that can be evaluated by furnishing their variables with definite values, a Boolean expression can be evaluated by giving Boolean values (i.e., 1 or 0) to its simple statements. The truth tables for the three basic logical operations 'and', 'or' and 'not' are:

a	b	$a \wedge b$
1	1	1
1	0	0
0	1	0
0	0	0

a	b	$a \vee b$
1	1	1
1	0	1
0	1	1
0	0	0

a	$\neg a$
1	0
0	1

A typical logic problem could be exemplified as follows: let $P(x)$ = 'x was French' and $Q(x)$ = 'x was a professional composer'. Find whether $P(x) \wedge Q(x)$ is true for the following values of x: Edward Elgar, Marie Curie, Camile Saint-Saëns. Thus:

> if $P(\text{Elgar}) = 0$ and $Q(\text{Elgar}) = 1$, then 0 and 1 = 0
> if $P(\text{Curie}) = 1$ and $Q(\text{Curie}) = 0$, then 1 and 0 = 0
> if $P(\text{Saint-Saëns}) = 1$ and $Q(\text{Saint-Saëns}) = 1$, then 1 and 1 = 1

The conclusion is that only Camile Saint-Saëns was both French and a professional composer.

Table 2.2

Commutation	$A \vee B = B \vee A$	$A \wedge B = B \wedge A$
Association	$A \vee (B \vee C) = (A \vee B) \vee C$	$A \wedge (B \wedge C) = (A \wedge B) \wedge C$
Distribution	$A \wedge (B \vee C) = (A \wedge B) \vee (A \wedge C)$	$A \vee (B \wedge C) = (A \vee B) \wedge (A \vee C)$
Complement	$A \vee \neg A = 1$	$A \wedge \neg A = 0$
Idempotency	$A \vee A = A$	$A \wedge A = A$
Identity	$A \vee 0 = A$	$A \wedge 1 = A$
Zero law	$A \vee 1 = 1$	$A \wedge 0 = 0$
de Morgan's	$\neg(A \vee B) = \neg A \wedge \neg B$	$\neg(A \wedge B) = \neg A \vee \neg B$
Involution	$\neg(\neg A) = A$	

The algebraic properties previously defined for set theory also apply to Boolean algebra. In this case, the symbol $\wedge$ plays the role of $\cap$, the symbol $\vee$ plays the role of $\cup$, the universally true statement plays the role of the universal set and a statement that is always false corresponds to the empty set (see Table 2.2).

As an example of the application of Boolean algebra to solve problems, imagine the data base of customers of an Internet-based record shop. The customers are classified in the data base according to their musical tastes:

A = Classical
B = Dance music
C = World music
D = Jazz
E = Rock
F = Miscellaneous

These categories may overlap in the sense that a single customer may like both classical music and jazz, for example. Here, $A(x)$ denotes the predicate 'x likes classical music', and so on. Thus, $A(\text{Suzana}) = 1$ means Suzana likes classical music and $E(\text{Luiz Carlos}) = 0$ means Luiz Carlos does not like rock. Now, suppose that the firm wishes to compile a list of people who like classical and world music but do not like jazz. This can be expressed as follows: $A(x) \wedge C(x) \wedge \neg D(x)$. Database search engines normally support queries in the form of Boolean expressions such as these; in this case, only those customers for which the whole expression is true will be added to the list. Essentially this is how companies select groups of customers satisfying combinations of criteria from a database in order to target their marketing.

The OpenMusic package provides some facilities to apply logical operations on musical data. Please refer to Chapter 8 and to OpenMusic's reference manual available on the CD-ROM for more details.

2.4 Introduction to matrices

A matrix is an arrangement of numbers in rows and columns. For example:

$$\begin{bmatrix} 2 & 0 \\ -7 & 4 \\ 1 & 87 \\ -3 & 4 \end{bmatrix}$$

This example is a 4×2 matrix in the sense that it has 4 rows and 2 columns. More generally, an $r \times c$ matrix has r rows and c columns.

An individual entry in a matrix is technically referred to as an *element* and it is represented in the form a_{rc}, that is, the element of the rth row and the cth column. For example, the element a_{32} in the above matrix is the number 87. Matrices are normally represented by capital letters and their elements with lower cases: $A = [a_{rc}]$.

Two matrices $A = [a_{rc}]$ and $B = [b_{rc}]$ are only equal if they are of the same dimension and if $a_{rc} = b_{rc}$ for all values of r and c. A matrix with the same number of rows and columns is a *square matrix* and the sequence of elements running from the top left to the bottom right of a square matrix is called the *diagonal* of the matrix. If there is a mirror-image of patterns of numbers reflected in the diagonal, than the matrix is said to be *symmetric*. Example:

$$\begin{bmatrix} 1 & 0 & 1 & 1 \\ 0 & 1 & 0 & 0 \\ 0 & 0 & 1 & 0 \\ 1 & 1 & 0 & 1 \end{bmatrix}$$

Sometimes, the rows and columns of a matrix need to be interchanged. This operation is called *matrix transposition*; for example $A = [a_{ij}]$ can be transposed to $A' = [a_{ji}]$ as follows:

$$A = \begin{bmatrix} 2 & 4 & 6 \\ 1 & 3 & 5 \end{bmatrix} \Rightarrow A' = \begin{bmatrix} 2 & 1 \\ 4 & 3 \\ 6 & 5 \end{bmatrix}$$

The power of matrices for representing and solving problems resides in the fact the we can apply arithmetical operations on them. In order to illustrate the summation of matrices, let us return to the example of the record shop. Suppose that the data base holds the information that one of the customers, Alex, has bought five CDs of classical music, six CDs of world music and ten miscellaneous CDs; so far she has not bought CDs of the other styles. This can be written as a one dimensional matrix, as follows: $S = [5\ 0\ 6\ 0\ 0\ 10]$. Note that the order of the 6-tuple follows the order $A, B, C, \ldots$ listed earlier. Next, suppose that Alex buys a further CD of classical music and two CDs of world music. The new acquisitions are represented as $N = [1\ 0\ 2\ 0\ 0\ 0]$. In order to add this new information to the date base, one simply needs to add the two matrices, element by element: $[6\ 0\ 8\ 0\ 0\ 10]$. Addition of matrices is therefore defined on an element-by-element basis with the restriction that only matrices of the same dimensions can be added. Another example:

35

$$\begin{bmatrix} 2 & 45 & -8 \\ 1 & 42 & 0 \\ 23 & 45 & 32 \end{bmatrix} + \begin{bmatrix} 65 & 4 & 8 \\ 0 & -32 & 2 \\ 7 & 5 & 21 \end{bmatrix} = \begin{bmatrix} 67 & 49 & 0 \\ 1 & 10 & 2 \\ 30 & 50 & 53 \end{bmatrix}$$

Matrices can be multiplied by a single number, or *scalar* as follows:

$$\begin{bmatrix} 2 & 45 & -8 \\ 1 & 42 & 0 \\ 23 & 45 & 32 \end{bmatrix} \times 2 = \begin{bmatrix} 4 & 90 & 16 \\ 2 & 84 & 0 \\ 46 & 90 & 64 \end{bmatrix}$$

For didactic purposes, the multiplication of one one-dimensional matrix by another could be defined in a similar way to addition on an element-by-element basis (Vince and Morris, 1990). For example, suppose that one wishes to calculate how much Alex has spent in the record shop. The prices are indicated in the second matrix, from top to bottom, according to the order *A, B, C,* etc. For example, the price for a CD of classical music is £5, for a CD of dance music is £6, and so on:

$$[6 \quad 0 \quad 8 \quad 0 \quad 0 \quad 10] \times \begin{bmatrix} 5 \\ 6 \\ 4 \\ 5 \\ 6 \\ 5 \end{bmatrix} = [30 \quad 0 \quad 32 \quad 0 \quad 0 \quad 50] \Rightarrow [112]$$

The method is rather straightforward: Alex has bought six CDs of classical music at £5 each, eight CDs of world music at £4 each, and so forth. The total amounts to £112. This operation also works for larger dimensions, with the condition that the matrices should be *compatible:* that is, the number of rows in the second matrix must match the number of columns in the first. For example, if we consider another customer, Kate, the operation would look like this:

$$\begin{bmatrix} 6 & 0 & 8 & 0 & 0 & 10 \\ 0 & 12 & 0 & 0 & 2 & 0 \end{bmatrix} \times \begin{bmatrix} 5 \\ 6 \\ 4 \\ 5 \\ 6 \\ 5 \end{bmatrix} \Rightarrow \begin{bmatrix} 112 \\ 84 \end{bmatrix}$$

Skipping the intermediate calculus, the result is: Alex spent £112 and Kate, who bought twelve CDs of dance music and two CDs of rock, spent £84. More generally, any two matrices are compatible for multiplication if the first has the same number of columns as the second has rows.

Finally, let us now consider the case where both matrices consist of more than one column and row. Suppose that the shop is planning Christmas sales and wishes to examine what their customers' expenditure would be if prices were reduced by a certain amount; the reduced prices are represented in the second column:

$$\begin{bmatrix} 6 & 0 & 8 & 0 & 0 & 10 \\ 0 & 12 & 0 & 0 & 2 & 0 \end{bmatrix} \times \begin{bmatrix} 5 & 3 \\ 6 & 3 \\ 4 & 2 \\ 5 & 3 \\ 6 & 4 \\ 5 & 2 \end{bmatrix} \Rightarrow \begin{bmatrix} 112 & 54 \\ 84 & 44 \end{bmatrix}$$

In this case, the arithmetic operation is exactly the same as in the earlier example, treating each column of the second matrix as if it were a separate matrix.

Matrices can also deal with elements other than numbers. It is possible, for example, to define matrices whose elements are Boolean values that function according to the logical operations introduced earlier. For example, it is possible to define a matrix of logical products as follows. Consider:

$$\begin{bmatrix} 1 & 1 \\ 0 & 0 \end{bmatrix} \wedge \begin{bmatrix} 1 & 1 \\ 0 & 1 \end{bmatrix}$$

The process is analogous to ordinary matrix multiplication where the logical operators $\wedge$ and $\vee$ play the role of the multiplication and addition, respectively. Thus:

$$\begin{bmatrix} 1 & 1 \\ 0 & 0 \end{bmatrix} \wedge \begin{bmatrix} 1 & 1 \\ 0 & 1 \end{bmatrix} = \begin{bmatrix} (1 \wedge 1) \vee (1 \wedge 0) & (1 \wedge 1) \vee (1 \wedge 1) \\ (0 \wedge 1) \vee (0 \wedge 0) & (0 \wedge 1) \vee (0 \wedge 1) \end{bmatrix} \Rightarrow \begin{bmatrix} 1 & 1 \\ 0 & 0 \end{bmatrix}$$

The final result comes from the simplification of the product matrix, according to the algebraic logical rules introduced in Section 2.3 above.

2.5 The basics of formal grammars

A computer can only make sense of the alpha-numeric symbols used by programmers because of the underlying formal grammar of the programming language being used. A formal grammar is a collection of either or both descriptive or prescriptive rules for analysing or generating sequences of symbols. Those streams of symbols that do not comply with the grammar of the language in question are not recognised by the machine. In human languages and music the picture is similar; in these cases the symbols are not alpha-numeric keystrokes but the words of a lexicon or musical parameters, such as notes and their attributes.

The examples in this section uses the following convention:

1) Capital letters such as A, B, C, etc. indicate *non-terminal nodes* (that is, non-terminal symbols of the grammar)
2) Lower cases such as a, b, c, etc. indicate *terminal nodes* (that is, terminal symbols of the grammar)
3) The symbol $\rightarrow$ stands for 'is defined as'
4) The symbol $|$ stands for 'or'

An example of a simple grammar G_1 might be defined as follows:

$$S \rightarrow A \mid B$$
$$A \rightarrow aA \mid a$$
$$B \rightarrow bB \mid b$$

In plain English, this simple grammar states that from S one can derive either A or B. If one derives A, then one can derive the string aA, by selecting the first option of the derivation rule, followed by the string aaA (by substituting A by aA), and so on, until the second option of the rule is chosen for deriving the terminal node a from A (by substituting A by a). Similarly, if one starts by deriving B from S, then one can derive from B any string b^k in k steps. The convention $\alpha \rightarrow \sigma$ is used to mean that a string σ can be derived from a string α, by applying one of the derivation rules of the grammar. Considering the grammar G_1, one could obtain the following sequence:

$$S \Rightarrow A \Rightarrow aA \Rightarrow aaA \Rightarrow aaaA \Rightarrow aaaaA \Rightarrow aaaaaA \Rightarrow aaaaaa$$

This sequence may also be represented visually in the form of a tree, as shown in Figure 2.9. It is often more convenient to use a superscript number to indicate length of a string or sub-string of identical terminal nodes; for example, the expression a^3b^4 is equivalent to the string $aaabbbb$.

As far as musical composition is concerned, grammars have great potential for defining musical production rules. The general notion of production rules based upon the notion of grammars can be formalised as follows: a grammar is a four-element structure (N, T, P, S) where:

1) N is a set of non-terminal nodes
2) T is a set of terminal nodes
3) P is a set of production rules in the form of $\alpha \rightarrow \sigma$
4) S is a symbol designated as the starting symbol of the grammar

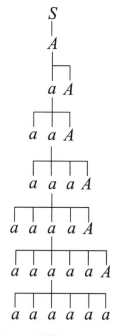

Figure 2.9 The representation, in the form of a tree, of a sequence derived from the grammar G_1.

As an example, let us define G_2 as $(\{S, A, B\}, \{a, b\}, P, S\}$ where P consists of the following production rules:

$S \;\rightarrow aSBC \mid aBC$
$CB \rightarrow BC$
$aB \rightarrow ab$
$bB \rightarrow bb$
$bC \rightarrow bc$
$cC \rightarrow cc$

Notice that in this case not all components of the left-hand side of the rules are single non-terminals. An example of a sequence generated by G_2 is given as follows:

$S \;\Rightarrow a\underline{S}BC$
   $\Rightarrow a\underline{aB}CBC$ (by using the rule $\underline{S} \rightarrow \mathbf{aBC}$)
   $\Rightarrow aab\underline{CB}C$ (by using the rule $\underline{aB} \rightarrow \mathbf{ab}$)
   $\Rightarrow aab\underline{B}CC$ (by using the rule $\underline{CB} \rightarrow \mathbf{BC}$)
   $\Rightarrow aabb\underline{C}C$ (by using the rule $\underline{bB} \rightarrow \mathbf{bb}$)
   $\Rightarrow aabb\underline{cC}$ (by using the rule $\underline{bC} \rightarrow \mathbf{bc}$)
   $\Rightarrow aabbcc$ (by using the rule $\underline{cC} \rightarrow \mathbf{cc}$)

For didactic purposes, the underlined elements are the ones that will be replaced by the elements in bold in the next step: $\underline{S}$ in $aSBC$ is replaced by $\mathbf{aBC}$, then $\underline{aB}$ in $aaBCBC$ is replaced by $\mathbf{ab}$, and so forth.

Grammars such as these are fundamental in computer science to establish whether a string is a valid within a domain. This is the core technique used by computers to parse and recognise the symbols of a programming language and also indicate when something is not syntactically correct. For instance, the example above indicates that the string *aabbcc* would be a valid statement for language G_2, because it is possible to trace it back up to its root S.

Grammars in which all production rules contain a single non-terminal node at the left-hand side are referred to as *context-free grammars* (CFG): let $G_3 = (N, T, P, S)$ so that every production in P is of the form $\Delta \rightarrow \sigma$ where $\Delta \in N$. It is called 'free' because the expansion will always apply, no matter what strings surround Δ.

2.5.1 Regular grammars and finite state automata

If a grammar $G_4 = (N, T, P, S)$ has a certain production rule of the form $S \rightarrow \sigma$ and S does not appear as a substituting element of the right-hand side of any other production in P, then G_4 is a *regular grammar*. An additional condition is that all other production rules of a regular grammar must be in either of the following two forms:

$A \rightarrow a$, where $A \in N$ and $a \in T$

$A \rightarrow aB$, where $\{A, B\} \in N$ and $a \in T$

Regular grammars may be represented visually using graph diagrams with labelled edges and vertices. In this case, there are distinct vertices in the graph, each labelled with an element of N, and one specially designed halting node labelled #. As a matter of convention, the vertex S (i.e., the starting node) is highlighted by an incoming arrow and the form of the halting vertex is a square, as opposed to a circle. Each edge of the graph represents a production rule. For instance, the production rule $A \rightarrow aB$ would be represented by a vertex A linked by an edge a to a B (Figure 2.10, left); in the case of $A \rightarrow a$, the edge a would link A to a terminal node # (Figure 2.10, right).

Figure 2.10 Two examples of graphs, each representing a single rule.

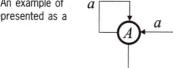

Figure 2.11 portrays the graph representation of the (regular) grammar G_5:

$S \rightarrow aA \mid bB$

$A \rightarrow aA \mid a$

$B \rightarrow bB \mid b$

Figure 2.11 An example of a grammar represented as a graph.

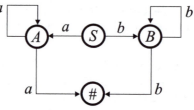

By tracing different paths in Figure 2.11, from the starting node S until reaching the terminal node #, one can form a variety of strings. For example, the path from S to B, to B again and then to #, produces $S \Rightarrow bB \Rightarrow bbB \Rightarrow bbb$. The set of all strings that can be obtained by tracing the paths of G_5 constitutes the *regular language* for G_5, represented as $L(G_5) = \{a^m \mid m > 1\} \cup \{b^n \mid n > 1\}$.

Note that regular grammars might be defined with any node of N designated as the starting node, and one or more nodes of N

designated as final nodes, as opposed to the standard S and $\#$ terminal nodes. In this case, they form a special case of regular grammars called *finite state automata* (FSA) or *finite state machines*. The formal definition of a FSA differs slightly from the definition of a grammar in the sense that S is no longer present; instead, there are $n_1 \in N$ for defining the starting node and $F \subset N$ to define the ending node. An example of a FSA is portrayed in Figure 2.12: $(\{A, B, C, D, E\}, \{a, b\}, P, A, \{D, E\})$, where

- $\{A, B, C, D, E\}$ is a set of non-terminal rules
- $\{a, b\}$ is a set of terminal modes
- P is a set of production rules (these can be inferred from the graph)
- A is the starting mode
- $\{D, E\}$ is a set of ending modes

Figure 2.12 An example of a finite state automaton.

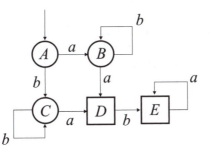

2.6 Brief introduction to probabilities

The topic of probability is popularly associated with gambling and weather forecasts. Simple probability problems are indeed related to dice-rolling and coin-tossing predictions and composers have always been interested in using the die for composing music; Mozart's musical dice game for composing waltzes comes to mind here (Schwanauer and Levitt, 1993).

As an example, consider the prediction for scoring more than four with a single throw of a die. Technically speaking, the throw of a die constitutes a *trial*. A trial is a process which can be expected to produce one outcome from the set of all possible outcomes; these outcomes cannot occur simultaneously. In the case of the die, the set of all possible outcomes from the die-rolling is $S = \{1, 2, 3, 4, 5, 6\}$. Our task is to predict when the result will be either five or six; this is normally referred to as the set of events $E = \{5, 6\}$. A further aspect to consider in a prediction is whether the trial is *fair*. That is, whether or not the probability of obtaining any one outcome is the same as for all other outcomes. In the case of the die, the trial is fair but there are many cases in which some outcomes are favoured over others (this will be discussed further in Chapter 3).

The chances of obtaining either the number five or six are two in six: the ratio of the number of the outcomes one is interested in obtaining, $E = \{5, 6\}$ and the total number of possible outcomes $S = \{1, 2, 3, 4, 5, 6\}$. This is formally written as:

$$p(e) = \frac{|E|}{|S|}$$

That is, provided that all the outcomes are equally likely, the probability of the occurrence of an event e is given by the ratio between the number of elements in the sub-set E by the number of elements in the set S:

$$p(5) = \frac{|2|}{|6|} = 0.33$$

The result of the above equation will always be a figure between 0 and 1. Thus, in order to estimate the probability that an event will not occur, one simply subtracts the probability of occurrence from one:

$$p(\neg e) = 1 - p(e)$$

2.7 A primer in computer programming

In principle, computers can be programmed to perform or solve almost any imaginable task or problem, with the proviso that the method for resolution can be explicitly defined and the data needed for its realisation can be explicitly represented. In essence, in order to program a computer one needs to write a sequence of instructions specifying how the machine will achieve the required results. This implies that the programmer must know how to resolve the problem or task in order to instruct the machine.

In many ways, a computer program is fairly similar to a recipe for a dish: a recipe gives clear steps to prepare the dish and lists all necessary ingredients. Recipes as well as computer programs must be void of ambiguities; e.g., the statement 'cover and simmer for five minutes over a low heat' is preferable to 'cover and simmer for some time'. Note, however, that even though the first statement is more precise than the second, the former still implies some background knowledge to interpret it. The term 'low heat', for instance, may carry different meanings to a chef, to a chemist and to an astrophysicist; the chef will probably know best how to select the gas mark of the cooker. People tend to develop specific coding and languages to communicate ideas on a specific subject more efficiently. The BBC Radio 4 Shipping Forecast is a good example of an efficient coding system to communicate the weather forecast to sailors clearly.

A programmer communicates instructions to the computer by means of a programming language. The idiosyncrasies of certain tasks require specific features which may require the design of specific programming languages to deal with them. Many programming languages have been developed for a diversity of purposes and types of computers, for example, C, Basic, Fortran, Java, Lisp, Miranda, Pascal and Prolog, to cite but a few. For example, whilst Prolog was specially devised for Artificial Intelligence research, Java was devised for applications involving the Internet. The main advantage of using a language that has been specially designed for specific problems is that its vocabulary normally employs the jargon of the domain applications.

Programs must be converted into binary machine code before the computer can actually perform any processing. Machine code is the only language that computers can actually understand but it is extremely tedious, if not impossible, to write a large complex program using raw binary code; hence the rationale for the so called high-level programming languages. Such programming languages provide a translator that converts the encoded program into machine code. There are two kinds of translators: *compiler* and *interpreter*. The compiler converts the whole code before executing it. Compilers save the result onto an executable file which can be activated at any time without having to repeat the translation process. Conversely, the interpreter converts each statement of the program and executes it before it proceeds to convert the next statement. Each mode has its own merits and pitfalls.

There are a number of programming languages specifically designed for music. They can be grouped into three categories, according to the abstraction boundary for which they were primarily designed (refer to Chapter 1): *languages for sound synthesis, languages for algorithmic composition* and *hybrid languages*. Languages for sound synthesis work at the microscopic level of music. We cite CLM and pcmusic as two significant examples for this group; note that they have already been introduced in another book in the Music Technology series by Focal Press (Miranda 1998). Conversely, languages for composition are aimed at the note level and upwards. Here we cite Common Music (Taube 1997), Tabula Vigilans (Orton and Kirk 1992), DMIX (Oppenheim 1994) and Formula (Anderson and Kuivila 1992). As for the remaining category, hybrid languages are music programming languages aimed at all levels of composition, ranging from synthesis to highly abstract musical structures. The most significant example of a hybrid language is Nyquist (Dannenberg 1997), which is available on the accompanying CD-ROM and is discussed in Chapter 8.

Regardless of the programming language or system used, one of the most fundamental programming skills to be acquired by anyone wishing to program a computer is the art of making algorithms. An algorithm is a sequence of instructions carried out to perform a task or to solve a problem. The sequence of rules defined earlier in section 2.1.2 to define a route in the graph is a fine example of an algorithm, but it is not well written. A better way of writing it would look like this:

```
BEGIN find_route algorithm
    1. Start with an empty set
    2. Select an edge
    3. IF the edge has not been selected before
        3.1 AND it has the lowest weight
        3.2 AND it does not form a closed path with the other
            edges already in the set
        3.3 THEN keep the edge
        3.4 ELSE discard the edge and proceed to step 4
    4. IF the set does not contain all of the vertices
        4.1 THEN go back to step 2
        4.2 ELSE proceed to END
END of find_route algorithm
```

Software engineers distinguish between an algorithm and a program. An algorithm is an abstract idea, a schematic solution or method, which does not necessarily depend upon a programming language. A program is an algorithm expressed in a form suitable for execution by a computer; it is the concrete realisation of the algorithm for use with a computer.

Depending on the complexity of the task or problem, a program may require a myriad of tangled, complex algorithms. In order to aid the design of neat and concise algorithms, the software engineering community has developed a variety of programming schemes and abstract constructs, many of which are now embedded in most languages, including those used for music composition. The more widespread of these programming schemes include: *encapsulated subroutines*, *path selection* and *iteration*, *passing data between subroutines* and *data structures*.

2.7.1 Encapsulated subroutines

One of the most fundamental practices of computer programming consists of organising the program into a collection of smaller sub-programs, generally referred to as *subroutines* (also known as *macro-modules, procedures* or *functions*). In this way, the same subroutine can be used more than once in a program without the need for rewriting. Subroutines may be stored in different files, frequently referred to as *libraries*; in this case, the compiler or interpreter must be able to retrieve subroutines from these libraries in order to merge them with the current program.

Most programming languages provide a library with a large number of subroutines which facilitate the ease of the programming task enormously. Programmers are actively encouraged to create their own library of functions. On highly developed programming desktops, the creation of a new program may simply require the specification of a list of subroutine calls.

In the example below, *algorithm A* is composed of a sequence of 4 instructions:

    BEGIN algorithm A
        Instruction 1
        Instruction 2
        Instruction 3
        Instruction 4
    END algorithm A

Suppose that this algorithm performs a task that will be requested several times by various sections of a larger program. Instead of writing the whole sequence of instructions again, the algorithm could be encapsulated into a subroutine called *Procedure A*, for example. This subroutine can now be invoked by other algorithms as many times as necessary. Example:

    BEGIN algorithm B
        Instruction 5
        Procedure A
        Instruction 6
    END algorithm B

The method or command for encapsulating a subroutine may vary enormously from language to language, but the idea is essentially the same in each case. As far as music processing is concerned, programming languages provide a library of ready-made subroutines which are the building blocks used to assemble synthesis and/or composition programs.

Roughly speaking, you could consider that an 'Instruction' is a basic command of the programming language for performing a simple operation such as printing a text on the screen, playing a MIDI note or calculating the square root of a number. A 'Procedure' could then be considered as a subroutine built on top of the commands, either provided as part of the programming language package or built by the programmer him/herself.

2.7.2 Path selection

The instructions and subroutines are performed in sequence by the computer and in the same order as they were specified. There are cases, however, in which an algorithm may have to select an execution path from a number of options, as demonstrated in the 'find route' example. The most basic construct for path selection is the *if-then* construct. The following example illustrates the functioning of the *if-then* construct. Note how *Instruction 2* is executed only if 'something' is true:

```
BEGIN algorithm A
      Instruction 1
      IF something
            THEN Instruction 2
      Instruction 3
      Instruction 4
END algorithm A
```

Another construct for path selection is the *if-then-else* construct. The *if-then-else* construct is used when the algorithm must select one of two different paths. Example:

```
BEGIN algorithm B
      Instruction 1
      IF something
            THEN Instruction 2
            ELSE Instruction 3
      Instruction 4
      Instruction 5
END algorithm B
```

In this case, if 'something' is true, then the algorithm executes *Instruction 2* and immediately jumps to execute *Instruction 5*. If 'something' is not true, then the algorithm skips *Instruction 2* and executes *Instruction 3* and *Instruction 4*, followed by *Instruction 5*.

The *if-then* and the *if-then-else* constructs are also useful in situations where the algorithm needs to select one of a number of alternatives. Example:

```
BEGIN algorithm C
      Instruction 1
      IF something
            THEN Instruction 2
```

```
        IF another thing
                THEN Instruction 3
                ELSE Instruction 4
        IF some condition
                THEN Instruction 5
        Instruction 6
    END algorithm C
```

Finally, there is the *case* construct. This construct is used when the selection of a path depends upon the multiple possibilities of one single test. In the example below, there are three 'possibilities' (X, Y and Z) that satisfy 'something'. Each 'possibility' triggers the execution of a different *Instruction*:

```
    BEGIN algorithm D
        Instruction 1
        CASE something
                possibility X THEN perform Instruction 2
                possibility Y THEN perform Instruction 3
                possibility Z THEN perform Instruction 4
        Instruction 5
    END algorithm D
```

In this example, *algorithm D* executes either *Instruction 2*, *Instruction 3* or *Instruction 4* depending on the assessment of 'something'; only one of these three instructions will be executed.

2.7.3 Iteration

Iteration (also referred to as a loop) allows for the repetition of a section of the program a number of times. There are many ways to set up an iteration, but all of them invariably need the specification of either or both the amount of repetitions or a condition to terminate the iteration. The most common constructs for iteration are *do-until*, *while-do* and *for-to-do*.

The following example illustrates the *do-until* construct:

```
    BEGIN algorithm A
        Instruction 1
        DO   Instruction 2
                Instruction 3
                Instruction 4
        UNTIL something
        Instruction 5
    END algorithm A
```

In this case, the algorithm executes *Instruction 1* and then repeats the sequence *Instruction 2*, *Instruction 3* and *Instruction 4* until 'something' is true. A variation of the above algorithm could be specified using the *while-do* construct, as follows:

```
BEGIN algorithm B
    Instruction 1
    WHILE something
    DO  Instruction 2
        Instruction 3
        Instruction 4
        Instruction 5
END algorithm B
```

The later algorithm (*algorithm B*) also repeats the sequence *Instruction 2*, *Instruction 3* and *Instruction 4*, but it behaves slightly differently from *algorithm A*. Whereas *algorithm A* will execute the sequence of instructions at least once, indifferent to the status of 'something', *algorithm B* will execute the sequence only if 'something' is true right at the start of the iteration.

The *for-to-do* construct works similarly to the *while-do* construct. Both may present some minor differences depending on the programming language at hand. In general, the *for-to-do* construct is used when an initial state steps towards a different state. The repetition continues until the new state is eventually reached. Example:

```
BEGIN algorithm C
    Instruction 1
    FOR initial state TO another state
    DO  Instruction 2
        Instruction 3
        Instruction 4
        Instruction 5
END algorithm A
```

In this case, the sequence *Instruction 2*, *Instruction 3* and *Instruction 4* will repeat until the state reaches 'another state'.

2.7.4 Passing data between subroutines

In order to construct modular programs consisting of a number of interconnected stand-alone subroutines, there must be a way to exchange information between them. As a matter of course, most subroutines (or functions) in a program need to receive data for processing and should pass on the results to other subroutines. Consider this example again:

```
BEGIN algorithm B
    Instruction 5
```

Procedure A
Instruction 6
END algorithm B

Note that *Algorithm B* calls *Procedure A* to perform a task and they are most likely to need to pass data to one another. The general scheme passing data that is commonly found in programming languages is illustrated below:

```
BEGIN fool_maths
    x = 2 + 3
    y = hidden_task(x)
    z = y + 5
END fool_maths
```

The *fool maths* algorithm is of the same form as *algorithm B*. After calculating a value for *x*, this value is passed on to a procedure called *hidden task*, which in turn performs a certain operation on *x*. The result of this operation is then given to *y*. Next, *y* is used to calculate the value of *z*.

The subroutine for the hidden operation must be defined elsewhere, otherwise *fool maths* will never work:

```
BEGIN hidden_task(a)
        b = a × 3
END hidden_task(b)
```

The role of *hidden task* is to multiply any value that is given to it. The variable that is inside the parenthesis on the right-hand side of the name of the subroutine represents the datum that is being passed and it is technically called the *parameter*. Note that the labels for the parameters do not need to match because the subroutines *fool maths* and *hidden task* are completely different from each other. However, the type (whether the parameter is a number, a word or a musical note, for example), the amount of parameters, and the order in which they are specified inside the parenthesis must match.

As for passing back the result, the variable on the left-hand side of the subroutine call automatically receives the result. The type of this variable must match with the type of result produced by the subroutine; in this case *y* (in *fool maths*) and *b* (in *hidden task*) match their types because they both are integer numbers. The result produced by the *fool maths* algorithm is 20 (Figure 2.13).

Figure 2.13 Complex algorithms are often composed of various subroutines that pass data to each other.

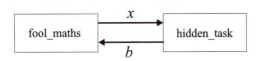

This is a simplified introduction to the passing data scheme, but it is all you need to know in order to follow the examples in the remaining chapters of this book. Bear in mind, however, that the actual implementation of this scheme may vary enormously, depending on the programming language used.

2.7.5 Data structures

The discrete mathematical concepts introduced above are fundamental for representing data structures on the computer. In order to write algorithms that deal with data structures such as sets, matrices, lists and so on, it is often necessary to indicate clearly how these entities should be processed. The programming scheme normally used for representing and manipulating data structures in an algorithm is the *array*; almost every programming language provides ways of implementing arrays.

An array can be one-dimensional (e.g. to contain lists of things), or multidimensional, to contain matrices. An array is normally represented by a capital letter and its elements by the lower case counterpart, furnished with an index. This index indicates the position of the element in the array. For example, an array A of five elements is represented as $A = [a(1), a(2), a(3), a(4), a(5)]$. In practice, our previous set of composers could be stored in an array as follows: C = [Antunes, Arcela, Carvalho, Garcia, Iazzetta, Lintz-Maues, Richter, Silva]. In this case, the element $c(3)$ is Carvalho and Richter is $c(7)$.

In the case of a multidimensional array (e.g., to contain a matrix), the elements have one index for each dimension. For instance:

$$A = \begin{bmatrix} a(1,1) & a(1,2) \\ a(2,1) & a(2,2) \end{bmatrix}$$

Thus, the number 112 in the following array N corresponds to the element $n(1,1)$ and $n(2,2) = 44$, to cite but two examples:

$$N = \begin{bmatrix} 112 & 54 \\ 84 & 44 \end{bmatrix}$$

In essence, arrays can be metaphorically associated with a large library bookshelf. Each book on the shelf has a code, or index. In order to retrieve a book from the shelf, librarians normally use a code that provides the co-ordinates to locate the book.

The following example illustrates how to fit an array into an algorithm. The algorithm below retrieves each element stored in an array of numbers, performs a multiplication on it and prints the result:

```
BEGIN multiply_elements
    V[n] = [12, 6, 4, 8, 2]
    FOR x = 1 TO 5
        DO   q = v[x]
                r = q × 10
                print(r)
END multiply_elements
```

The array *V* has 5 elements. Each element of the array can be retrieved by an indexed variable. For example $v(1) = 12$, $v(2) = 5$, $v(3) = 4$, and so on. The *for-to-do* iteration scheme is used to repeat a sequence of instructions (for multiplication and for printing) five times. At each iteration of the *for-to-do* loop, the value of x is incremental and serves as an index to retrieve the values from the array. For instance, at the third iteration $x = 3$ and therefore $v(3) = 4$, which in turn is passed to the variable q. Next, r is the result of the multiplication of q by 10; this result is printed. It is assumed that the *print(r)* instruction is a subroutine defined elsewhere. When the five iterations are completed, the outcome of the algorithm will be the following values printed (either on the screen of the computer or on paper): 120, 60, 40, 80, 20.

2.7.6 A musical example

As an example, let us study an algorithm that includes the programming schemes introduced above:

```
;
; This algorithm produces a simple melody
;
BEGIN produce_melody
    N[n] = [60, 62, 58, 56, 66, 65, 67]
    init = N[1]
    play(init)
    FOR x = 1 TO 7
        DO m = N[x]
            IF m < 60
                THEN p = transp(m)
                play(p)
                ELSE play(m)
    play(init)
END produce_melody
;
```

First of all, note that the lines beginning with a semi-colon are regarded as comments and are not part of the algorithm as such. Programmers are always encouraged to document their codes with comments in plain natural language in order to facilitate maintenance and to aid the interpretation of the code by other programmers that may eventually need to understand it. Almost all programming languages allow for the inclusion of comments. The array N contains a set of musical notes represented as MIDI numbers (Rumsey 1994). The task of the algorithm is to pick notes from this set and play them in a certain sequence (Figure 2.14). There is one main condition to be met: if the note picked is lower that 60 (i.e., middle C) then it has to be transposed one octave higher. To begin with, the first note of the array, $N(1) = 60$, is retrieved and stored in a variable called *init*; then this note is played. It is assumed that *init* will hold the value 60 in memory until it is explicitly told otherwise. Then the iteration begins: the value of x will increase from 1 to 7 and at each time step the algorithm performs a number of things. Firstly, it retrieves the note respective to the current index value (e.g., if $x = 2$, then $N(x) = 62$) and gives it to variable m. Then it checks whether this note is lower than 60. If so, then m is sent to a subroutine called *transp*, which will return the transposed value of m to a variable p. Next, the note p is played and the algorithm proceeds to the next step of the iteration.

```
;————————————————————————————————————
; Subroutine for transposing a note
; one octave upwards
;————————————————————————————————————
BEGIN transp(a)
; this subroutine transposes one octave up
    b = a + 12
END transp(b)
;————————————————————————————————————
```

If m is not lower than 60, then the algorithm skips the transposition task and goes on to play m unchanged. When the *for-to-do* interaction reaches the seventh step, then it breaks the loop, plays the *init* note again and ends the algorithm. The algorithm should by then have played the following sequence: 60, 60, 62, 70, 68, 66, 65, 67, 60. A similar algorithm could have been written to deal with the duration of the notes, their intensity and so forth (Figure 2.14).

Figure 2.14 The algorithm picks notes from the set (refer to main text) in order to play a melody.

2.8 The legacy of The Second Viennese School

In the early 1920s, a composer in the midst of the German-oriented contemporary music scene, eager for a reasonable way forward through the rather chaotic state of the atonal music trend of the beginning of the twentieth century, came up with a method for musical composition of historical importance: his name was Arnold Schoenberg and the method was called *serialism*.

Schoenberg proposed that the standard twelve musical notes should be treated by the composer as having equal importance in relation to each other, as opposed to the hierarchical organisation of tonal musical scales, where some notes are considered to be more important than others. In practice, he wanted to abolish the idea of using tonal scales (C major, C minor, etc.) by favouring the idea of a series of twelve notes organised according to principles other than the ones given by classical harmony. This may sound obvious to a twenty-first century musician, but Schoenberg's ideas were revolutionary at the time. As there is an immense body of musicological research about the implications of Schoenberg's work for current musical praxis, this section will focus only on the practical aspects of his compositional method.

A series of twelve notes functions here as the blueprint for the entire composition. Basically there was only one law to be strictly observed in Schoenberg's method: each note of the series should be used in the same order in which they occur in the series, all way through the piece; furthermore, there can be no repetition of a note before all the notes of the series have been used. The particular sequence of notes that forms a series is entirely arbitrary; Figure 2.15 illustrates an example of a series.

Figure 2.15 An example of a 12-tone series.

It is not difficult to realise that a composition where the series is repeated over and over might become dull after a few moments. Of course Schoenberg was aware of this and cleverly allowed for some transformations on the twelve notes of the series: the series could be played *forwards*, *backwards* (or retrograde), *upside-down* (Figure 2.16) or *backwards upside-down* (Figure 2.17). Coincidentally or not, this was the very same type of treatment that medieval composers applied to musical modes in order to make their compositions more interesting; medieval composers would not have dreamt of changing tonality (sic) within the same piece, mostly because this would sound bad, as medieval instruments were not tuned for this. Furthermore, Schoenberg's method also made use of transposition; i.e., the series could start on a different note, provided it maintained the same sequence of intervals between the notes.

Figure 2.16 In music, playing something upside down is technically called 'inversion'. A clock face is a good example to illustrate inversion: if the original interval moves forward eight notes on the clock, from two to ten, then the inversion moves backwards eight notes from two to six.

Figure 2.17 The four versions of the series portrayed in Figure 2.15.

Forwards

Backwards

Upside-down

Backwards upside-down

In short, a twelve-tone piece can be thought of as a perpetual variation on the series using a few prescribed transformational mechanisms. As it stands, Schoenberg's work in itself is not of much interest here, but rather the results of the work of his pupils and other sympathisers of the serial method, most notably the ideas put forward by Anton Webern, Olivier Messiaen and latterly by Pierre Boulez (Boulez 1963).

Originally, Schoenberg proposed serialism as method for organising only notes; to a large extent, other aspects of his pieces followed standard practices and forms. Webern, however, introduced other organisational principles in his music and prepared the ground for Messiaen, Boulez and others to extend the serial method technique to organise musical parameters other than notes; such as duration, dynamics and modes of attack. In the early 1950s Boulez wrote a piece for two pianos called *Structures* which is entirely serial. The serial processes defined by Boulez could generate the piece almost automatically: the piece is in perpetual transformation and no pitch recurs with the same duration, dynamic or mode of attack. In addition to the pitch series, he defined a duration series, a dynamics series (from *pppp* to *ffff*) and a modes of attack (*staccato*, *mezzo staccato*, etc.) series. The serial processes determined the overall structure of the composition, ordering the different series in a combinatorial-like fashion.

Mozart, Beethoven and many other great composers of classical music, had lived in Vienna. Schoenberg, Webern and Alban Berg, another important pupil of Schoenberg, were also based in Vienna: hence the reason they are often referred to as the Second Viennese School. They fostered the creation of a musical language in which every element influencing the content of a musical expression was controlled by explicit and formal rules: pitch, duration, attack and dynamics. A completely serial piece of music such as *Structures* will never perhaps be appreciated by a wide audience, but nevertheless the legacy of using formal and explicit rules to generate pieces of music is undeniably important for anyone intending to use computers to compose music. *Structures* is the paramount example of the parametrical thinking discussed in Chapter 1.

A few algorithms for serialism are given below. The first algorithm is a straightforward subroutine for transposing a twelve-tone series. The notes are represented as MIDI numbers in an array.

```
;————————————————————————————
; Transpose a given series of notes by a
; given amount of semitones
;————————————————————————————
BEGIN transpose(A[n], amount)
    B[n] = create_empty_array(12)
    FOR x = 1 TO 12
        DO B[x] = A[x] + amount
END transpose(B[x])
;————————————————————————————
```

The input for the algorithm is an array $A(n)$ containing the twelve notes, and the amount to be transposed in terms of semitones. Firstly, the algorithm calls a subroutine to create an empty array $B(n)$ of a size equal to 12; this array will contain the transposed notes. Then, the elements of the array *A(n)* are retrieved one at a time, transposed and placed in array *B(n)*. Once the loop terminates, the algorithm returns the array *B(n)* containing the transposed notes. For example, if one calls transpose([62, 70, 68, 66, 65, 60, 58, 72, 70, 62, 70, 72], 5), then the result will be [67, 75, 73, 71, 70, 65, 63, 77, 75, 67, 75, 77].

The following algorithm implements the retrograde process:

```
;————————————————————————————————
; Generate the retrograded version of
; a given series of notes
;————————————————————————————————
BEGIN retrograde(A[n])
    B[n] = create_empty_array(12)
    y = 1
    FOR x = 12 TO 1
            DO B[y] = A[x]
            y = y + 1
    END retrograde(B[y])
;————————————————————————————————
```

The mechanism is similar to the one used by the previous algorithm, with the difference that the loop runs from 12 to 1 because we need to read array *A[n]* backwards. The index for array *B* is the variable *y*, which functions as a counter: it increases its value at each loop-step. For example, when *x* = 12 then *y* = 1, when *x* = 11 then *y* = 2, and so on.

In order to generate the upside-down version of a series, first of all one must calculate the distance between the current note and the previous one. The upside-down note is produced by subtracting this distance value from the previous note.

```
;————————————————————————————————
; Generate the upside-down version
; of a given series of notes
;————————————————————————————————
BEGIN inversion(A[n])
    B[n] = create_empty_array(12)
    B[1] = A[1]
    FOR x = 2 TO 12
            DO  y = A[x] – A[x–1]
                B[x] = A[x–1] – y
    END inversion(B[x])
;————————————————————————————————
```

In order to test the algorithm manually, let us input the following for inversion: *A[n]* = [60, 62, 56, ...]

Step 1: inversion([60, 62, 58, ...]).
Step 2: empty algorithm B[n] is created

Step 3: the first element of the inverted series is the same as the original, that is $B[1] = 60$

Step 4: the *for-to-do* loop begins; $x = 2$

Step 5: the distance between the current note $A[2] = 62$ and the previous one $A[1] = 60$ is calculated, that is $y = 2$

Step 6: the inversion of the current note is calculated as $B[2] = B[1] - y$, that is $B[2] = 58$

Step 7: second step of the loop; $x = 3$

Step 8: the distance is calculated $y = 56 - 62$, then $y = -6$

Step 9: the inversion is calculated $B[3] = 58 - (-6)$, then $B[3] = 64$

...

...

So far the result of the inversion is $B[n] = [60, 58, 64, ...]$.

Finally, the algorithm for generating the backwards upside-down version of the series can simply call on the two previous subroutines:

```
;
; Generate the retrograded upside-down
; version of a given series of notes
;
BEGIN retrograde_inversion (A[n])
    B[n] = create_empty_array(12)
    C[n] = create_empty_array(12)
    B[n] = retrograde(A[n])
    C[n] = inversion(B[n])
END retrograde-inversion(C[n])
;
```

The OpenMusic tutorial on the CD-ROM provides practical examples of the serial operations discussed in this section. Please refer to Chapter 8 and to OpenMusic's tutorials number five and six.

2.9 The legacy of formalised music

The radical serialism fostered by Pierre Boulez and others in Europe naturally had many opponents. John Cage in the USA is a notorious example: rather than tie his pieces to a serial kernel, Cage preferred to toss a coin or play ancient divination games for generating musical material (Revill 1992). However opposed, both trends have led to new ways of thinking about music, which emphasises the generative processes that underlie the origins of a piece, be they random or highly controlled.

Another composer, Iannis Xenakis, proposed an interesting approach to composition, which is neither strictly serial nor loosely playful. Xenakis has inspired many of today's computer musicians:

his approach embodied statistical principles. Scientists use statistics to make general predictions concerning a mass of random fluctuations, such as the overall speed and trajectory of the molecules in a cloud of gas. Xenakis purported that a similar rationale could be applied to large numbers of musical events, such as the sound of a mass of strings in different glissando ranges.

Although Xenakis is commonly cited as being a pioneer in using probabilities for composing music, the aspect of his work that interests us most here is the abstract formalism that underlines his musical thought (probability in music is discussed in Chapter 3). The following example embodies a musical formalism based on set theory, inspired by the theoretical framework introduced by Xenakis in his book *Formalized Music* (Xenakis, 1971).

Let us define three sets of natural numbers, *P*, *D* and *I*, corresponding to pitch intervals, duration and intensity, respectively; these sets are ordered.

A general law of composition for each of these sets may be established as follows: Let v_m be a vector of three components, p_n, d_n and i_n, such that $p_n \in P$, $d_n \in D$ and $i_n \in I$, respectively, arranged in this order $v_m = \{p_n, d_n, i_n\}$. In this context, a vector is a point in a three-dimensional space, whose co-ordinates are given by p_n, d_n and i_n (Figure 2.18). The particular case of the vector in which all the components are zero is referred to as the zero vector, v_0, that is, the origin of the co-ordinates.

Figure 2.18 A vector represents a point in a multidimensional space.

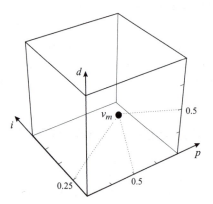

Considering the algebraic properties of sets introduced earlier, let us define two compositional laws for vector v_m: the *additive law* and the *multiplication law*.

The additive law:

$$(v_i = \{p_m, d_n, i_o\}) + (v_j = \{p_x, d_y, i_z\}) \Rightarrow (v_i + v_j) = \{(p_m + p_x), (d_n + d_y) \, (i_o + i_z)\}$$

Multiplication law (c is a natural number):

$$c \times (v_i = \{p_m, d_n, i_o\}) \Rightarrow c \times v_i = \{c \times p_m, c \times d_n, c \times i_o\}$$

An additional constraint to be observed is that p_n, d_n and i_n should not take arbitrary values, but values within the audible range. For the purpose of this example, the default vector for the origins of the co-ordinates is defined as $v_0 = \{60, 4, 64\}$. The elements of set P are represented in terms of MIDI-note semitones, the elements of set D are represented in terms of metric beats and the elements of I in proportional terms from *pianissimo* to *fortissimo* whose values range from 0 to 127. The origins of the co-ordinates are therefore: the note C3, a duration of one beat and an intensity equal to *mezzo piano*.

We are now in a position to define some generative rules for generating musical events. The following statements define three vector generators; np means n units above the origin of p whereas $-np$ means n units below the origin, and so on:

$$v_1 = \{14p, 4d, i\}$$
$$v_2 = \{(X + 12)p, Yd, Zi\}$$
$$v_3 = \{(X - 4)p, Yd, Zi\}$$

The F_n statements below define the formation of sequences of vectors:

$$F_1 = [v_1]$$
$$F_2 = [v_2, v_2, 2 \times v_2]$$
$$F_3 = [v_3, v_1 + v_3, v_3]$$

As implemented on a computer, suppose that each F_n statement has been implemented as functions that generate the vectors; in this case F_2 and F_3 receive the list of parameters for v. This could be denoted as $F_n(x, y, z)$, where x, y and z are lists of values for the variables of v.

A generative score could then be given to the system which in turn would generate the passage portrayed in Figure 2.19, as follows:

Figure 2.19 A musical passage automatically generated by the generative musical score.

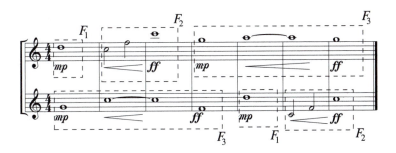

59

Top stave: $\{F_1, F_2[(0, 2, 32), (5, 2, 48), (0, 2, 32)], F_3[(24, 4, 0), (12, 4, 48), (24, 4, 64)]\}$

Bottom stave: $\{F_3[(12, 4, 0), (0, 4, 48), (10, 4, 64)], F_1, F_2[(-12, 2, 32), (-7, 2, 48), (0, 2, 32)]\}$

Note that F_1 does not need any input parameter because v_1 needs no variable to function; v_1 always produces the same note: $\{14p, 4d, i\}$.

3 Probabilities, grammars and automata

3.1 Probabilities

The idea of devising systems for composing music using randomness has always fascinated composers, especially after the invention of the computer. But the raw outcome of a random process seldom fulfils of the expectations of exigent composers: they often prefer to assess its suitability for inclusion in their music according to their aesthetic preferences and musical aims. The conditions under which the randomly generated elements are used are therefore crucial for a composition.

The reason that randomness has gained so much popularity in the computer age is probably because computers can actually do better than just throwing a die or tossing a coin: they can be programmed to perform testing on the randomly produced material in order to assess its suitability for the piece in question. Moreover, as an alternative to the random-and-test process, musicians can design programs that embed compositional constraints within the random process itself. One way to do this is to use probabilities.

Chapter 2 introduced the fundamentals of probabilities and the concept of a *fair trial*, that is, the case in which the chances for obtaining a specific outcome are the same for all possible outcomes. There is no favouritism in a fair trial; for example, whilst the prediction of the outcome from rolling a die is fair, the same is not true for a football match between a strong and a weak team. Note that in this case the information about the past performance of the two teams is taken into account in order to infer which is the stronger team, whereas the former case does not depend upon any information about the previous outcomes of the rolled dice. Uncertain processes whose outcome is influenced by previous results are referred to as *conditional probabilities*.

In music, probabilities are normally used to generate musical materials by picking up elements from a set. For example, in order to generate a melody out of a set of notes, the computer could be programmed to randomly select one note at a time and then play it through a synthesiser. If there are no repeated notes in the set, then the trial is fair. Conversely, if there is one or more duplicated note in the set, then the chances of this particular note being selected for play will increase. Intuitively, this is how probabilities operate: by repeating notes in a set, the chances of those repeated notes being picked will increase proportionally to the number of repetitions. Mathematics, however, provides tools for making probability-based selections other than selection by merely repeating the elements in a set: *distribution functions* are the epitome of such tools. In this context, the term *stochastic generator* denotes a system that generates sequences of musical parameters by selecting them from a given set according to some specified distribution function.

3.1.1 Distribution functions

Originally, distribution functions were made to express the probabilities of a continuous random phenomenon, but they also apply to discrete contexts. Basically, there are five classes of distribution functions *uniform, linear, exponential, concave* and *convex* distributions.

The *uniform distribution* is the simplest type of all five basic distribution functions: it embodies the fair trial case where the

Figure 3.1 The graph for the uniform distribution.

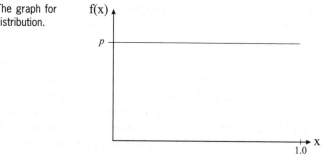

chances of an outcome are equal for all the other outcomes. The graph for the uniform distribution is illustrated in Figure 3.1: all outcomes will be within the range between zero and one (on the horizontal axis) and the flatness of the plotted line indicates that there is an equal chance for all values within this range.

Note that the probability is expressed in terms of the likelihood that a result will fall within the region of possible outcomes. In practice, the probability is given by the area delimited by the horizontal line at the top (technically called a *curve*), the lowest number range on the left and the highest number range on the right. For example, in Figure 3.2 the probability that the result will be a number between 0.1 and 0.2 is a 10% chance (0.1×1.0), whereas the probability that the result will be a number between 0.5 and 0.8 is 30%. The algorithm for generating uniformly distributed values uses only a single random numbers generator to produce values between 0.0 and 1.0:

```
;
; Generator of uniformly distributed values
;

BEGIN uniform
    uni = random(1.0)
END uniform(uni)
;
```

Figure 3.2 A probability function expresses the likelihood that a result will fall within a region of possible outcomes.

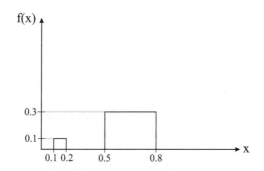

Figure 3.3 portrays the case of *linear distribution*, whereby the values on the left-hand side of the horizontal axis have a higher probability of outcome than the ones on the right-hand side. The linear distribution could also be inverted in order to decrease the probability values between the two limits. By combining sequences of linear distributions one can create break-point linear distributions. As far as computer implementation is concerned, in order to obtain the linear distribution shown in Figure 3.3 the algorithm could simply generate two uniformly distributed random numbers between zero and one, and then select the smaller one for the result:

```
;—————————————————————————————————
; Generator of linearly distributed values
;—————————————————————————————————
BEGIN linear
    x = random(1.0)
    y = random(1.0)
    IF x > y
            THEN lin = x
            ELSE lin = y
END linear(lin)
;—————————————————————————————————
```

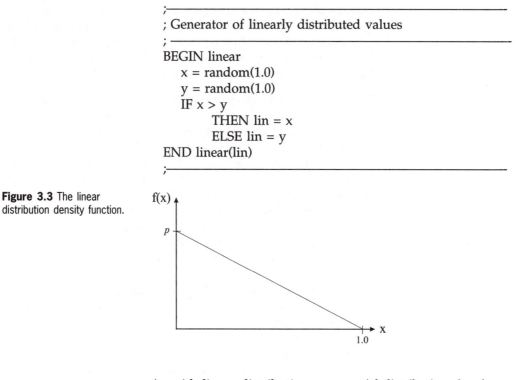

Figure 3.3 The linear distribution density function.

As with linear distribution, *exponential distribution* also favours the left hand values on the axis, with the difference that there is a parameter λ that controls the degree of this favouritism; the larger the value for λ the greater the tendency to obtain the left hand values (Figure 3.4). A simple method for generating exponentially distributed values can be defined as follows: a) produce a uniformly distributed value greater than zero, but not higher than one, b) divide this value by λ and c) compute the natural logarithm of the result:

```
;—————————————————————————————————
; Exponentially distributed values
;—————————————————————————————————
BEGIN exponential(lambda)
    x = random(1.0)
    y = x / lambda
    exp = logarithm(y)
END exponential(exp)
;—————————————————————————————————
```

Concave distribution exhibits a bilateral exponential distribution. Its density function has the shape of a concave curve and it is also characterised by a parameter λ where, in this case, the larger the value of λ, the wider the curve. Here the outcome is more

Figure 3.4 The exponential
distribution function.

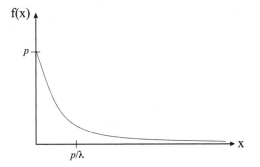

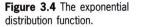

Figure 3.4 The exponential
distribution function.

likely to fall within the central values of the horizontal axis of
the graph (Figure 3.5). A concave function can be approximated
by taking the tangent of a scaled uniformly distributed value
equal to or higher than 0.0, but less than 1.0; the scale factor is
$\pi/2$ (i.e., 1.570796). The concave value is obtained by scaling the
result of the tangent to the λ parameter. The λ parameter is given
as input to the algorithm:

```
;
; Concave distribution approximation
;

BEGIN concave(lambda)
    x = random(0.999)
    s = x × 1.570796
    t = tangent(s)
    conc = t × lambda
END concave(conc)
;
```

Concave distributions are those that seem to best model the
occurrence of natural phenomena in the real world. They are
essentially symmetrical because they are centred on a mean
value. Concave distributions whose values are skewed rather
than centred on a mean value are called asymmetrical and they
also are useful for musical applications.

Figure 3.5 The symmetrical
concave distribution graph.

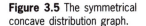

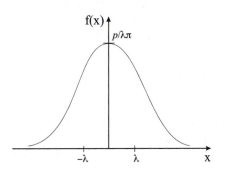

Finally, *convex distributions* have the inverse shape of concave ones in the sense that their peak values are at the endpoints. As with concave distributions, convex distributions can also be either symmetrical or asymmetrical. An example algorithm for producing values within the convex distribution curve is given as follows:

```
;————————————————————————————————
; Concave distribution approximation
;————————————————————————————————
BEGIN convex(x y)
    invx = 1 / x
    invy = 1 / y
    rand1 = 0
    rand2 = 0
    sum = 2.0
    WHILE sum > 1.0
        DO   WHILE rand1 = 0
                DO rand1 = random(1.0)
             WHILE rand2 = 0
                DO rand2 = random(1.0)
             a = rand1 ∧ invx
             b = rand2 ∧ invy
             sum = a + b
    conv = a / sum
END convex(conv)
;————————————————————————————————
```

The symbol ∧ denotes an exponential; for example *rand1∧invx* is equivalent to $rand1^{invx}$. There are two input parameters for the procedure: x and y. Firstly, the algorithm calculates the inverse of these parameters (e.g., if $x = 0.5$ then the inverse is $1/0.5 = 2$) and initialises the values for the variables *rand1*, *rand2* and *sum*. Next, it produces two uniformly distributed values higher than zero but not higher than one. These two values are elevated to the exponential of the inverse values of x and y (i.e. *invx* and *invy*, respectively) and the results are added. The result of this addition must not be higher than one, otherwise these values are discarded and the algorithm repeats the process, various times if necessary, until a result that satisfies this condition is obtained.

3.2 Probability tables

The previous section explained how a variety of probabilistic processes can be described by different distribution functions and gave an example of how to generate numbers pertaining to a certain distribution curve. Depending upon the application, a distribution function can be efficiently implemented as a probability table that holds values corresponding to the likelihood of the occurrence of one or more events. In a compositional system, probability tables may be utilised as part of a decision making routine. In the case of the uniform distribution, the task of selecting one of x uniformly distributed events can be implemented simply by selecting a random number between 0 and $x - 1$, which in turn is used to call the sub-routine that is associated with it:

```
;────────────────────────────────────────
; Select one of four uniformly distributed operations
; and apply it to a given array of notes
;────────────────────────────────────────
BEGIN uniform_table(A[n])
    B[n] = create_empty_array(12)
    rand = random(3)
    CASE
        rand = 0 THEN B[n] = transpose(A[n], 5)
        rand = 1 THEN B[n] = transpose(A[n], 7)
        rand = 2 THEN B[n] = retrograde(A[n])
        rand = 3 THEN B[n] = inversion(A[n])
END uniform_table(B[n])
;────────────────────────────────────────
```

The uniform-table procedure receives one input which is an array $A[n]$ of n notes. Then it randomly selects one of four possible operations to perform on these notes and returns the result of the operation to an array $B[n]$.

In order to implement probability tables for distributions other than the uniform one, we simply substitute the random numbers generator by a subroutine that generates numbers pertaining to the respective distribution curve, as illustrated in Section 3.1.

Another, and perhaps more efficient, method for implementing probability tables uses the *cumulative mechanism*. The best way to understand how this mechanism works is to imagine a set of points that divide a line going from 0.0 to 1.0 into sub-segments whose widths correspond to the different probabilities of the individual events (Figure 3.7). In order to select an outcome, we randomly choose a number between 0.0 and 1.0 and examine which segment it falls into. The sum of all the values of a probability table must be equal to 1.0. This condition guarantees that some outcome will always take place when the table is called upon.

In a compositional system, the composer normally specifies a number n of possible outcomes, each of which corresponds to:

1) a procedure for generating specific musical events if that particular outcome is selected and 2) a number between 0.0 and 1.0 that gives its probability of being selected (Figure 3.6). In practice, these probability values correspond to partitions of a line segment from 0.0 to 1.0 into n distinct sub-segments, or regions, whose interval widths correspond to the probabilities of the respective outcomes (Figure 3.7). Then, a random number between 0.0 and 1.0 is generated and the decision is made by observing in which part of the n segments the number lies.

Figure 3.6 Procedures for generating specific musical events are associated to probability coefficients.

Procedure	P
produce_melody	p_1
transpose(A[n], amount)	p_2
retrograde(A[n])	p_3
inversion(A[n])	p_4
retrograde_inversion (A[n])	p_5

Figure 3.7 The probability values correspond to partitions of a line segment that goes from 0.0 to 1.0 into n distinct sub-segments.

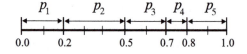

The algorithm for the cumulative mechanism is given as follows:

```
;————————————————————————————
; Generate a random number between 0 and 1. Case it
; lies in the segment which corresponds to the probability
; of the first event then execute the respective procedure.
; Case it does not lie in the first segment but in the second
; segment, then execute the second procedure, and so on.
;————————————————————————————
BEGIN cumulative_choice(P[i], A[n])
    B[n] = create_empty_array(12)
    rand = random(1.0)
    CASE
        rand < P[0]
            THEN B[n] = transpose(A[n], 5)
        rand < P[0] + P[1]
            THEN B[n] = transpose(A[n], 7)
        rand < P[0] + P[1] + P[2]
            THEN B[n] = retrograde(A[n])
        rand < P[0] + P[1] + P[2] + P[3]
            THEN B[n] = inversion(A[n])
END cumulative_choice(B[n])
;————————————————————————————
```

The *cumulative choice* procedure receives two inputs: an array of probabilities and an array of notes. One of four possible operations will be applied to the array of notes $A[n]$ and the result of the operation is returned to the array $B[n]$. Each outcome has a unique procedure associated with it. The probability values are stored in the array $P[n]$, where n is an index for retrieving the actual value in the array. For example, if array $P = [0.5, 0.18, 0.22, 0.1]$, then $P[1] = 0.5$, $P[2] = 0.18$, and so on.

3.3 Markov chains

Markov chains are conditional probability systems where the probability of future events depends on one or more past events. The number of past events that are taken into consideration at each stage is known as the *order* of the chain; a Markov chain that takes only one predecessor into account is of first order, whereas a chain that considers both the predecessor and the predecessor's predecessor is of second order, and so on.

In general, an nth-order Markov chain is represented by a state transition matrix of $n + 1$ dimensions. The state transition matrix gives us information on the likelihood of an event's occurrence, given the previous n states. Figure 3.8 portrays an example of a state transition matrix for a first-order Markov chain with four possible outcomes: A, B, C and D. The past outcomes are listed vertically, and the current states are listed horizontally. In order to find the probability of outcome B occurring immediately after A, for example, one simply finds state A in the first column (in this case A is a past outcome), and then horizontally moves along to the B column. The entry in this case is 0.1; that is, there is a one in ten chance of an outcome B occurring after an outcome A.

Figure 3.8 An example of a first-order Markov chain.

	A	B	C	D
A	0.2	0.1	0.3	0.4
B	0.5	0.1	0.2	0.2
C	0.5	0.2	0.1	0.2
D	0.2	0.2	0.3	0.2

3.3.1 Mathematical properties of Markov chains

Markov chains have some interesting mathematical properties that are worth noting. For example, a state X is said to be *reachable* from state Y if is it possible to reach state X from state Y

after a finite number of steps. If state Y is reachable from state X and X is reachable from state Y, then the two states are said to *communicate*. For example, in Figure 3.8, state C is reachable from A and A is reachable from C, therefore A and C communicate.

The communication relation on a Markov chain constitutes an *equivalence* because it satisfies the following requirements:

- it is *reflexive*: a state always communicates with itself
- it is *symmetrical*: if a state X communicates with a state Y, then clearly Y communicates with X
- it is *transitive*: if a state X communicates with a state Y, and state Y communicates with a state Z, then state X also communicates with the state Z

The states of a Markov chain can be grouped into equivalent classes of communicating states. Those states that are certain to occur again once they have been reached by the chain are called *recurrent*, and the equivalence class they belong to is known as a *recurrent class*. States that may never occur again (i.e., those that are not recurrent) are called *transient*, and the class they belong to is known as a *transient class*. Every Markov chain should contain at least one recurrent class and possibly transient classes; a Markov chain cannot contain only transient classes.

A Markov chain containing exactly one recurrent class and possibly some transient classes is referred to as *ergodic*. Ergodic Markov chains are preferred for musical composition because they enable composers to make rough predictions about their behaviour and outcome (Grimmet and Stirzaker 1982).

3.3.2 Generating note streams

As an example to illustrate the use of a Markov chain to generate streams from a set of notes, consider the ordered set of 8 notes that constitute a C major scale as follows: {C3, D3, E3, F3, G3, A3, B3, C4} (Figure 3.9).

Figure 3.9 A C major scale notes set.

C3 D3 E3 F3 G3 A3 B3 C4

Next, assume the following rules for establishing which notes are allowed to follow a given note:

- if C3, then either C3, D3, E3, G3 or C4
- if D3, then either C3, E3 or G3
- if E3, then either D3 or F3
- if F3, then either C3, E3 or G3
- if G3, then either C3, F3, G3 or A3
- if A3, then B3
- if B3, then C4
- if C4, then either A3 or B3

Each of these rules represents the transition probabilities for the next note in a sequence. For example, after C3, each of the five notes C3, D3, E3, G3 and C4 has a 20% chance each of occurring; that is, each of these notes has the probability $p = 0.2$. Conversely, notes F3, A3 and B3 will never occur; that is, these notes have the probability $p = 0.0$. In this case the probability has been uniformly distributed between the five candidates, but this does not necessarily need to be so. For instance, one could establish that after D3, note C3 has the probability $p = 0.2$ of occurring whereas notes E3 and G3 have the probability $p = 0.4$ each.

The above rules can be expressed in terms of probability arrays. For instance, the probability array for note C3 is $p(C3) = [0.2, 0.2, 0.2, 0.0, 0.2, 0.0, 0.0, 0.2]$ and for node D3 is $p(D3) = [0.2, 0.0, 0.4, 0.0, 0.4, 0.0, 0.0, 0.0]$, and so on. The order of the probability coefficients in the array corresponds to the order of the elements in the set. The probability arrays for all the notes of C major can be arranged in a two-dimensional matrix, thus forming a Markov chain (Figure 3.10).

Figure 3.10 An example of a first-order Markov chain for the notes of a C major scale.

	C3	D3	E3	F3	G3	A3	B3	C4
C3	0.2	0.2	0.2	0.0	0.2	0.0	0.0	0.2
D3	0.33	0.0	0.33	0.0	0.33	0.0	0.0	0.0
E3	0.0	0.5	0.0	0.5	0.0	0.0	0.0	0.0
F3	0.33	0.0	0.33	0.0	0.33	0.0	0.0	0.0
G3	0.25	0.0	0.0	0.25	0.25	0.25	0.0	0.0
A3	0.0	0.0	0.0	0.0	0.0	0.0	1.0	0.0
B3	0.0	0.0	0.0	0.0	0.0	0.0	0.0	1.0
C4	0.0	0.0	0.0	0.0	0.0	0.5	0.5	0.0

Higher order Markov chains in which the next outcome depends on more than one generation of previous occurrences, work similarly to the first-order ones as introduced above, with the difference that the matrices will have more dimensions: one for each generation. For example, the matrix for a second-order chain must have three dimensions; one for the outcome in question, one for the last outcome and one for the second-to-last.

3.3.3 Random walk processes

A Markov chain whose matrix representation has non-zero entries immediately on either side of the main diagonal, and zeros everywhere else, constitutes a *random walk process*. Imagine a pet robot standing on a staircase on a step between the first step s_1 and the eighth step s_8, being able to move up or down one step at a time. If it has a probability p to go upwards, then it will have the probability $q = 1 - p$ to go downwards. An example of a matrix for a random walk process whereby the probabilities to go up or down are all equal, is given as:

$$W = \begin{bmatrix} 0.0 & 1.0 & 0.0 & 0.0 & 0.0 & 0.0 & 0.0 & 0.0 \\ 0.5 & 0.0 & 0.5 & 0.0 & 0.0 & 0.0 & 0.0 & 0.0 \\ 0.0 & 0.5 & 0.0 & 0.5 & 0.0 & 0.0 & 0.0 & 0.0 \\ 0.0 & 0.0 & 0.5 & 0.0 & 0.5 & 0.0 & 0.0 & 0.0 \\ 0.0 & 0.0 & 0.0 & 0.5 & 0.0 & 0.5 & 0.0 & 0.0 \\ 0.0 & 0.0 & 0.0 & 0.0 & 0.5 & 0.0 & 0.5 & 0.0 \\ 0.0 & 0.0 & 0.0 & 0.0 & 0.0 & 0.5 & 0.0 & 0.5 \\ 0.0 & 0.0 & 0.0 & 0.0 & 0.0 & 0.0 & 1.0 & 0.0 \end{bmatrix}$$

Random walk processes are useful for generating musical phenomena that require smooth gradual changes over the material. For instance, by substituting the matrix of the Markov chain of Figure 3.10 by the matrix W above one could obtain a sequence of notes of the type portrayed in Figure 3.11.

Figure 3.11 A sequence of notes generated by a random walk process.

3.4 Formal Grammars

A piece of music can be thought of as a tangled hierarchical structure: at the lowest level are the notes of the piece, which in turn form phrases and melodies, then themes, movements and so forth. An example of a hierarchical representation for a sonata-like form is partially shown in Figure 3.12. In this case, the entire piece fits nicely in a single large hierarchy of nested substructures, but as a matter of course, interesting musical pieces do not normally display such a neat structure. Notwithstanding, compositions do seem to have at least some type of layered organisation.

Figure 3.12 The hierarchical representation of a sonata-like form.

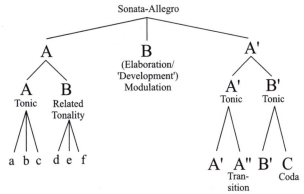

As discussed in Chapter 1, the hierarchical approach to musical thought has a striking similarity to linguistics, in the sense that musical notes can correspond to phonemes, which in turn form words, sentences, paragraphs, and so on. This similarity has naturally fostered a mutual interest between linguists and musicologists. In fact, it is true to conjecture that the origins of linguistics come from an essentially musical interest. One of the oldest linguistic treatises known to mankind was written in India by a scholar called Pānini; it dates back to over 25 centuries ago. The motivation for this treatise was the Vedic belief in the power of the Vedic hymns when the words were chanted properly. The objective to devise a formal tool to study the Sanskrit was to find ways to harness and preserve its mystical power. Pānini described the phonetic structure of Sanskrit and also devised strict rules for composing phonological structures. What is interesting here is that Pānini's treatise is not simply normative, but also generative: the rules were also used to generate well-formed Sanskrit utterances (Holtzman 1994).

By and large, linguistics is concerned with the make-up of the words and sentences of a language. A language here is formally defined as a subset of the infinite set formed by all possible combinations of a set of basic units. For example, musical notes can be considered as the units for a musical language. In this case, all possible combinations of notes constitute the universal set of all possible musical compositions within this domain. A musical language can therefore be thought of as being a subset of combinations dictated by certain constraints and rules. The following paragraphs focus on a particular formalism developed by linguists for formalising such constraints and rules, which can be of great use in musical composition using the computer: *formal grammars.*

The notion of formal grammars is one of the most popular, but also controversial, notions that has sprung from linguistics to fertilise the ground ready for computer music. Formal grammars appeared in the late 1950s when linguist Noam Chomsky published his revolutionary book *Syntactic Structures* (Chomsky 1957). To put it in simple terms, Chomsky suggested that humans are able to speak and understand a language mostly because we have the ability to master its grammar. According to Chomsky, the specification of a grammar must be based upon mathematical formalism in order to thoroughly describe its functioning; e.g., formal rules for description, generation and transformation of sentences. A grammar should then manage to characterise sentences objectively and without guesswork. In general terms Chomsky believed that it should be possible to define a universal grammar, applicable to all languages. He purported that a language can be studied from two distinct structural viewpoints labelled *deep structure* and *surface structure*. The former embodies the fundamental structure shared by all languages of the world. His thesis is that the human mind is hardwired with the faculty to process the deep structure at birth. The surface structure then embodies the particularities of specific languages that are acquired by children as they grow up within a particular culture.

Many composers working with computers have been strongly influenced by the work of Chomsky. Indeed, some musicologists and linguists believed that Chomsky's assumptions could be similarly applied to music (Lerdhal and Jackendoff 1983) (Cope 1987). A substantial amount of work inspired by the general principles of structural description of sentences has been produced, including a variety of useful formal approaches to musical analysis (Cook 1987). We are not in a position here to evaluate the plausibility of Chomsky's theories for linguistics and musicology. Rather, we are just interested in the fact that they can indeed be pragmatically used for compositional purposes.

One of the most successful applications of formal grammars for music composition is David Cope's EMI system. This particular system uses a technique for writing formal grammars known as an advanced transition network (ATN) combined with a pattern extraction mechanism in order to craft works in a particular style. The system takes as an input a few examples of pieces in the style that the user wishes to replicate. Then the system scans the input looking for short musical passages that indicate the style of the pieces. The system is then able to produce new passages incorporating variations on the previously scanned musical material, building up a complete composition, piece by piece, whilst ensuring that it fits the desired grammatical design. One could roughly say that Cope programmed the computer with the deep structure of musical capacity and the input examples shape the surface structure of the system. More information about EMI can be found in David Cope's own book *Computers and Musical Style* (Cope, 1991).

3.4.1 A brief introduction to formal grammars

The mathematical basis of formal grammars can be found in Chapter 2. Figure 3.13 illustrates an example of a grammatical rule for a simple affirmative sentence: 'A musician composes the music'.

Figure 3.13 Example of a grammatical rule.

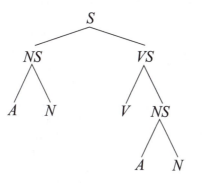

The above rule states that a sentence, represented as *S*, is formed by a noun-sentence represented as *NS*, and a verb-sentence, represented as, *VS*. A noun-sentence *NS* is formed by an article *A* and a noun *N*, and a verb-sentence *VS* is formed by a verb *V* and a noun-sentence *NS*. This could also be written as follows (note that the notation here is slightly different from the one in Chapter 2, in the sense that the symbol + for 'and' is used in place of the symbol | for 'or'):

$$S \rightarrow NS + VS$$
$$NS \rightarrow A + N$$
$$VS \rightarrow V + NS$$

Such a grammatical rule can be programmed onto a computer in order to generate sentences automatically. In this case, the computer also must be furnished with some sort of lexicon of words to choose from. For example:

$$A = \{the, a, an\}$$
$$N = \{dog, computer, music, musician, coffee\}$$
$$V = \{composes, makes, hears\}$$

Note that the lexicon defines the three classes of words required by the rule: articles, nouns and verbs, represented by the sets *A*, *N* and *V* respectively. Given the above rule and lexicon, the computer can be activated to generate meaningful sentences, such as 'A computer composes the music' or 'A dog hears the musician', but also nonsense ones, such as 'A musician composes the dog' or 'A coffee hears the computer'. The production of nonsense could be alleviated by tightening the grammar with more specific rules, but meaning is very hard to formalise. Fortunately, musicians should not need to worry about musical semantics. Meaning in music is a much harder issue to deal with than meaning in language, but most musicians would surely agree that it is preferable to leave this issue unsolved anyway. Here it suffices to assume that a syntactically well-formed musical string should sound good; whether it fits specific musical contexts is a matter of musical aims, aesthetic preferences and cultural convention.

The rule above is termed *generative* because it is used to generated sentences from scratch. Other types of grammatical rules are *transformational* rules for applying transformations onto existing sentences. In this case, the computer is programmed to verify if the sentence to be transformed is syntactically correct prior to the transformation. This verification is often done by matching the generative rule that would have generated the sentence. (This notion of 'transformational rules' is slightly different from the one posed by linguistics, but this is not significant here.) For example, a transformational rule to change the order of simple sentences could be defined as follows:

IF:
$$S(o) \rightarrow NS(n) + VS(m)$$
$$NS(n) \rightarrow A(n) + N(n)$$
$$VS(m) \rightarrow V + NS(m)$$

THEN:
$$S(t) \rightarrow NS(m) + VS(n)$$
$$NS(m) \rightarrow A(m) + N(m)$$
$$VS(n) \rightarrow NS(n) + V$$

In English, the above rule reads as follows: If the sentence to be transformed is composed of a first noun-sentence $NS(n)$ followed by a verb-sentence $VS(m)$, and the first noun-sentence is formed by an article $A(n)$ and a noun $N(n)$, and the verb-sentence is formed by a verb (V) and a second noun-sentence $NS(m)$ of the same format of the first noun sentence, then the transformed sentence will be formed by the second noun-sentence $NS(m)$ followed by a new verb-sentence composed of the first noun-sentence $NS(n)$ followed by the verb V. If this transformation rule is applied to the sentence 'A musician composes the music', then it will give the result 'The music a composer composes'. Punctuation could also be included in the rule in order to produce 'The music, a composer composes'.

Generative and transformational rules to generate and transform musical sentences (sic) can be similarly defined.

3.4.2 An example of a grammar for music composition

In order to define a grammar for music we should carefully consider what the constituents of the rules will be; e.g., notes, phrases, melodies, chords, etc. For the purposes of the example below, we defined the constituents of our grammar in terms of five fundamental notions:

- the notion R_n of a reference note (e.g., R_1 = C4)
- the notion of interval I_n between two notes (e.g., I_1 = perfect 5th)
- the notion of direction D_n of the interval (e.g., D_1 = upwards)
- the notion of sequence SEQ_n
- the notion of simultaneity SIM_n

An example of a generative rule for our grammar is defined as follows (Figure 3.14):

$$SIM_1 \rightarrow SEQ_1 + SEQ_2$$
$$SEQ_1 \rightarrow (I_5, D_1) + (I_8, D_1) + (I_{11}, D_1)$$
$$SEQ_2 \rightarrow (I_5, D_2) + (I_8, D_2)$$

Figure 3.14 An example of a musical grammar.

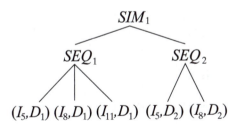

The lexicon of our grammar comprises two sets, one set of intervals I and one set of interval directions D, as follows:

I = {minor 2nd, major 2nd, minor 3rd, major 3rd, perfect 4th, augmented 4th, perfect 5th, minor 6th, major 6th, minor 7th, major 7th, octave, none}
D = {upwards, downwards, none}

The pair (I_5, D_1), for example, indicates that the interval is a perfect fourth (the fifth element of the set I) and the direction of the interval is upwards (the first element of the set D). Thus, the rule reads as follows: 'A certain musical passage is composed of two sequences played simultaneously. One sequence is formed by three notes and the other is formed by two notes. The notes of the former sequence are calculated from a given reference note in this order: a perfect fifth upwards, a minor sixth upwards and a major seventh upwards. The notes of the latter sequence are calculated, using the same reference note, in this order: a perfect fourth downwards and a minor sixth downwards'. By establishing that the reference point $R_1 = C4$, the rules could produce the musical passage shown in Figure 3.15:

Figure 3.15 A musical passage generated by the grammar.

In order to keep our example simple, other musical aspects, such as rhythm, have not been included in our grammar, but it would be perfectly feasible to take these into account as well.

A transformational rule, for the above type of passage could, for example, create a new sequence SEQ_3 by joining the two sequences into a single simultaneous event ($SIM_2 \rightarrow SEQ_1 + SEQ_2$), followed by the original first sequence SEQ_1 (Figure 3.16):

IF:
$$SIM_1 \rightarrow SEQ_1 + SEQ_2$$
$$SEQ_1 \rightarrow (I_5, D_1) + (I_8, D_1) + (I_{11}, D_1)$$
$$SEQ_2 \rightarrow (I_5, D_2) + (I_8, D_2)$$

THEN:
$$SEQ_3 \rightarrow SIM_2 + SEQ_1$$
$$SIM_2 \rightarrow SEQ_1 + SEQ_2$$

Figure 3.16 An example of a musical transformation.

The computer could be programmed to produce an entire musical composition by successive activation of a variety of generative and transformational rules (Figure 3.17):

Figure 3.17 Musical material produced by successive activations of rules.

3.5 Finite state automata

Finite state automata (FSA) are normally used as a tool for designing programming languages. They are equivalent to formal grammars in many respects. Whilst linguists would employ formal grammars for studying the underlying laws of the formation of the words and sentences of a language, a computer scientist would employ finite state automata to specify the laws of formation of strings for building parsers for the compilers or interpreters of a programming language. In general, the underlying principles of both are similar but the formalisms are different. As far as music is concerned, finite state automata tend to be more efficient than formal grammars for lower level symbolic constructions and short musical passages. Conversely, formal grammars seem to perform better than automata when it comes to higher level control of musical form.

A slight variation of the FSA definition introduced in Chapter 2 can be given as follows: $A = (Q, I, T, E)$, where:

1   Q is a set of elements called *states*
2   I is subset of Q whose elements are called *initial states*
3   T is a subset of Q whose elements are called *terminal states*
4   E is a set of compound elements called *edges*

The compound elements of E are formed by combining the elements of Q two by two, by means of a labelled link (Figure 3.18). For example, an element of E represented as (p, σ, q) is an edge from p to q linked by an operator σ. The basic idea of the FSA is that an edge of the kind (p, σ, q) transforms state p to state q under the action of σ. In this case, p is the *initial state* whilst q is the *terminal* state of the edge (p, σ, q).

Generally speaking, FSA define transformational paths from one state to another. More intricate paths from an initial state p to r in a certain finite state automaton may therefore involve a sequence of edges, $e_1, e_2, ..., e_n$, such that p is the initial state of e_1, r is the terminal state of e_n and the terminal state of intermediate edges is the initial state of the edge next to it. Figure 3.19 portrays a visual representation, technically referred to as a

Figure 3.18 A labelled link from element p to element q represents an edge.

Figure 3.19 A simple finite state automaton.

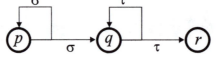

directed graph, of the following finite state automaton (note that in this example there are two operators for the edges: σ and τ):

$$A = (\{p, q, r\}, \{p\}, \{r\}, \{(p, \sigma, p), (p, \sigma, q), (q, \tau, q), (q, \tau, r)\})$$

Mathematicians use rather elegant formalisms to demonstrate whether or not a path is recognised by the automaton, but this is not necessary at this stage. As far as we are concerned, it suffices to say that given a finite state automaton, one can define a set of rules to produce paths that are recognisable to the automaton.

FSA have a great deal to offer the computer musician. Their graphical representation helps to make the underlying concepts easier for the non-mathematician to grasp and can be of great assistance for specifying the rules and when translating them into some programming language.

The simplest way of implementing finite state automata for a musical purpose is to label the edges of the automaton with notes or short musical events; for instance, in the example below, they correspond to short rhythmic cells. This affords the composer an efficient and highly visual method for producing musical sequences that adhere to a set of generative rules. As an example, consider the simple directed graph portrayed in Figure 3.20. This automaton can be used to produce rhythmic sequences of the forms illustrated in Figures 3.21, 3.22 and 3.23.

Figure 3.20 An example of a finite state automaton with the edges corresponding to rhythmic figures.

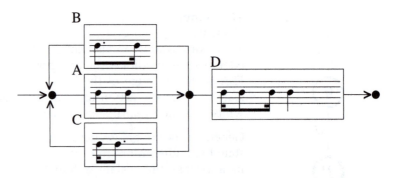

The interpretation of Figure 3.20 is as follows: a rhythmic sequence begins with an unspecified number of combinations of the rhythmic cells constructed by placing either element B or C

Figure 3.21 The shortest rhythmic sequence that can be generated by the automaton in Figure 3.20.

after element A. The sequence must end with element D. According to this automaton, the shortest rhythmic sequence that it can generate is the one composed of element A followed by element D (Figure 3.21). The next shortest rhythmic sequence can be either A followed by B followed by D or A followed by C followed by D (Figure 3.22), and so on (Figure 3.23).

Figure 3.22 An example of a sequence that is one 'edge' larger than the one in Figure 3.21.

Figure 3.23 Example of a large sequence generated by the automaton in Figure 3.20.

Clearly, the example generates an infinite number of different rhythms, although without much variety because the automaton is wholly deterministic. There are a few variants that could be made in the definition of the finite state automaton introduced above in order to make it more powerful. One variant could be specified by introducing a memory device so that the grammar could remember previously generated strings; this type of automata is known as the *pushdown automata* (Rayward-Smith 1983). Such automata would allow for the construction of strings which depend upon previously generated characters. Musically, this would correspond to, for example, a recapitulation section occurring only if it succeeded a development section, which in turn, had succeeded an exposition section.

3.6 Related software on the accompanying CD-ROM

Texture (in the folder **texture**) is geared to generating random MIDI messages but the user can draw distribution curves in order to bias the output. It is possible, for example, to specify that MIDI velocity values (i.e., the loudness of the notes) between 0 and 64 are twice as likely to occur as those between

65 and 127. Tangent and Koan are also chiefly based on probabilities for generating music. In fact, most software for generating music uses probabilities in one way or another.

An example of the application of Markov chains can be found
in CAMUS 3D (in the folder **camus**). Here Markov chains are
used to control rhythm and the temporal organisation of note
groups (see Figure 8.16 in Chapter 8).

As for grammars, there is Bol Processor BP2 (in the folder **bp2**).
The package comes with a number of examples of grammars for
generating music, ranging from Carnatic Indian music (e.g., file
-gr.trialMohanam) to minimalist music in the style of Steve Reich
(e.g., file -gr.NotReich).

4 Iterative algorithms: chaos and fractals

4.1 Iterative processes

An iterative process is the repeated application of a mathematical procedure where each step is applied to the output of the preceding step (Figure 4.1).

Figure 4.1 An iterative process whereby the output is fed back to the input.

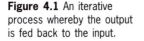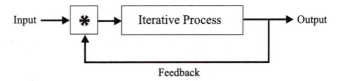

Input → **✳** → | Iterative Process | → Output

Feedback

Mathematically, an iterative process is defined as a rule that describes the action that is to be repeatedly applied to an initial value x_0. The outcome of an iterative process constitutes a set, technically referred to as the *orbit* of the process; the values of this set are referred to as the *points* of the orbit. Thus, the orbit O that arises from the iterated application of a rule F to an initial value x_0 is written as: $O^F(x_0)$. For example, consider the following rule F: $x_{n+1} = x_n + 2$. This rule indicates that the next value of the orbit x_{n+1} is calculated by adding the two units to the previous value. If one specifies that the initial value of x_0 is equal to zero, then the result of the iterated application of F onto x_0 will be $O^F(0) = \{0, 2, 4, 6, ...\}$. This is certainly a very simple orbit, but iterative processes have the potential to produce fascinating orbits, some of which can be used to generate interesting musical sequences. Essentially, an iterative process may produce three classes of orbits:

1   orbits whose points tend towards a stable fixed value
2   orbits whose points tend to oscillate between specific elements
3   orbits whose points fall into chaos

As an example of the first class of orbits, consider the rule $x_{n+1} = (x_n/2)$. If $x_0 = 1$, then the result of the iteration will be $O^F(1) = \{1, 0.5, 0.25, 0.125, ...\}$. In this case the orbit will invariably tend towards zero, no matter what the initial value is.

By way of an example for the oscillatory class of orbits, consider the following rule: $x_{n+1} = 3.1x_n(1 - x_n)$. If $x_0 = 0.5$, then the outcome of the iteration will be $O^F(0.5) = \{0.5, 0.775, 0.540, 0.770, 0.549, 0.768, 0.553, 0.766, 0.555, 0.766, 0.556, 0.765, 0.557, 0.765, 0.557, 0.765, ...\}$. Note that after an initial settling stage, the orbit falls into an oscillatory state between 0.765 and 0.557 (Figure 4.2). This is a rather simple oscillation of two periods, but oscillations may involve many more periods than this.

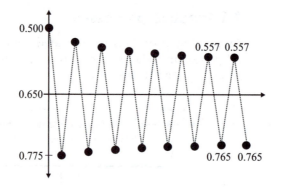

Figure 4.2 After an initial settling stage, an iterative process may fall into an oscillatory state.

In the third class of orbits, it is not possible to distinguish an explicitly recurrent pattern; hence they are referred to as *chaotic orbits*. Sometimes it is no mean thing to determine whether an iterative process has fallen into a complex oscillatory state with a large period or whether it has fallen into chaos. There are three basic principles that we can watch for in order to assess whether an iterative process is behaving chaotically or not: *high-sensitivity to initial conditions, period doubling process* and *sporadic settlements*.

The principle of high-sensitivity to initial conditions states that if tiny variations in the value of x_0 cause a significantly different outcome after a few iterations, then it is likely that the iterative process is behaving chaotically. This principle was nicknamed the *butterfly effect* by metereologist Edward Lorenz. Lorenz built a computer model of a hypothetical global weather system and observed that small changes to the initial variables of the model could lead to extremely diverse weather phenomena. He illustrated the highly sensitive nature of the global weather system by suggesting that even the flapping of a butterfly's wings on one side of the globe could be a possible cause for a hurricane on the other side (Ekerland, 1995).

As a simple practical example to illustrate the class of chaotic orbits, consider the rule $x_{n+1} = 3.1x_n(1 - x_n)$ given earlier, and substitute the constant value 3.1 by a variable Δ, as follows: $x_{n+1} = \Delta x_n(1 - x_n)$. Then, if we set $\Delta = 4$ and compare the orbits produced by two very close values for x_0, say 0.3 and 0.301, we can clearly see that the resulting orbits are very different from one another (Figure 4.3).

Figure 4.3 Two examples of an iterative process where significantly different orbits are produced as a consequence of very small variations in the initial value. The upper graph (a) portrays the outcome produced by setting the initial value to 0.3 and the lower graph (b) portrays the outcome produced by setting this value to 0.301.

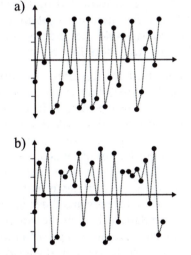

One should note, however, that although sensitivity to initial conditions is a characteristic of chaotic systems, systems that are sensitive to initial conditions, but that are not chaotic, do exist.

The principle of period doubling refers to the successive doubling of the number of different points in the orbit. This can be illustrated using the same equation as the previous example: by leaving the intial value x_0 constant and varying the value of Δ between three and four, we obtain the outcome shown in Figure 4.4. Note that as the value of Δ increases, the number of points duplicates from two, to four, then eight, sixteen and so on until a truly chaotic orbit is achieved, where $\Delta = 4$.

Figure 4.4 The period doubling effect. As the value of Δ increases, the number of points doubles.

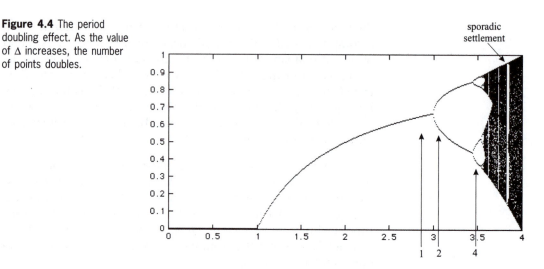

Finally, the principle of sporadic settlements can be observed by looking at the region indicated by the arrow at the top right of Figure 4.4. As the iterative system evolves towards chaotic behaviour, quasi-stable trajectories sporadically emerge, but soon fall into period doubling again.

The equation used for the above examples is called the *logistic equation* and it is a well-known model for population growth: $x_{n+1} = \Delta x_n(1 - x_n)$ where $0 \le \Delta \le 4$. In order to illustrate how this model works, suppose that a team of ecologists is monitoring the annual population growth of anaconda snakes in the Amazon forest. The population of anacondas in a certain year n is denoted as a fraction x_n of some maximum possible population that the forest can support, such that $0 \le x_n \le 1$. In this case one can say that $(1 - x_n)$ represents the largest possible fraction of the maximum population that can be added to the current population x_n. The factor Δ denotes a growth rate that embodies various factors imposed by the ecological constraints of the environment (e.g., prey resources, climatic conditions, deforestation, predatory threats, and so forth). Thus this equation could be read as: an estimation of the population of anacondas in the Amazon forest for next year is given by multiplying the growth rate by the present population and then by the amount of additional population than can be supported by the ecosystem of the forest.

Another famous example of an iteractive system that displays chaotic behaviour is the *Hénon attractor*, also known as the *strange attractor*. The Hénon attractor is a simple model devised in the early 1970s by the French astronomer, M. Hénon, to describe the trajectory of a floating object around a gravitational body. The model essentially has two equations as follows: $x_{n+1} =$

Figure 4.5 An example produced by the Hénon model.

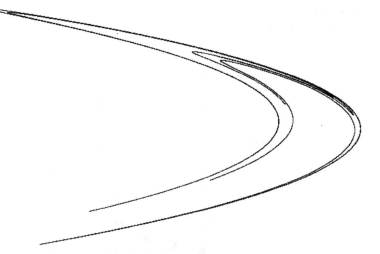

$y_n - \Phi x_n^2$ and $y_{n+1} = \Delta x_n$, where Φ and Δ are constant coefficients. In practice, at each iteration the equations generate the co-ordinates for plotting the points of the orbit in a graph. Figure 4.5 shows the orbit generated by the *Hénon attractor* when $\Phi = 1.4$ and $\Delta = 0.3$.

One of the most interesting iterative rules is, however, this one: $z_{n+1} = z_n^2 + c$. This rule actually defines a fractal called the Mandelbrot set and it is intriguing because depending upon the value of c, the points of the orbit may either grow indefinitely (e.g., $c = 1$) or remain bounded by a certain range of values (e.g., $c = -1.38$). (Fractals will be introduced later; basically, it is a geometry for describing irregular and fragmented structures.)

Figure 4.6 Two examples of orbits generated by the function $z_{n+1} = z_n^2 + c$. The top graph (a) portrays an oscillatory orbit generated by $c = -1.38$ and the bottom one portrays a chaotic orbit generated by $c = -1.9$.

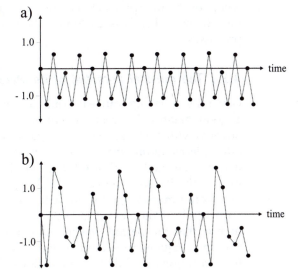

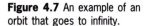

Figure 4.7 An example of an orbit that goes to infinity.

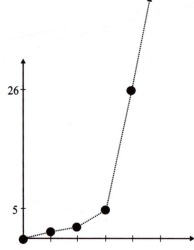

The graph in Figure 4.7 shows what happens when $z_{n+1} = z_n^2 + 1$. In this case the orbit is said to go towards infinity, as opposed to the orbits of both graphs shown in Figure 4.6, where the orbits remain bounded.

Although most fractals result from iterative functions, in the context of this book fractals will be introduced as an alternative to Euclidean geometry rather than as types of iterative systems that might or might not display chaotic behaviour. And indeed, it is possible to generate fractal structures by means of functions that are not iterative. Before we go on to study fractal geometry and the Mandelbrot set, it is opportune at this stage to set the scene for the interplay between iterative processes and music by introducing a brief discussion on the musical potential of these processes.

4.1.1 The musical potential of iterative processes

Generally speaking, our ear tends to enjoy music that presents a good balance between repetition of musical elements and novelty within the scope of the piece itself and in relation to other pieces. Orbits that quickly converge towards stable values may not be the best choice for generating music, as the behaviour of the iterative process quickly becomes static. Oscillatory orbits offer more scope for producing interesting musical results, particularly if the period is large. The only fly in the ointment here is that the period of such oscillatory orbits is often small and the frequent repetition of musical parameters can, with a few exceptions, rapidly become tedious. By far the most promising type of orbit, from a musical point of view, is the chaotic one. A chaotic orbit tends to wander through a fixed range of elements, visiting similar, though not identical points. This phenomenon has the potential to generate new musical materials with some degree of correlation to previous materials. Intuitively, we could think of generating music from a chaotic orbit as a compositional process akin to writing themes and variations, or the exposition and development of musical passages. However, finding an effective method for mapping the orbits onto musical parameters is not an easy task. This is one of the greatest difficulties that composers face when working with algorithmic composition systems that use the output from essentially non-musical processes; that is, non-musical in the sense that they were not originally developed with a musical perspective in mind. It should be noted, however, that composers of the past also faced such compositional burdens in one way or another. For instance, by the time the fugue was conceived as a way to structure musical form, rules for constructing good fugues simply did not exist; it took years of experimentation until reasonable methods could be settled. Similarly, we do not yet have clear rules for constructing chaotic musical forms.

The correlation between iterative processes and music is, to a certain extent, dictated by the dimensionality of the iterative process at hand. For example, a two-dimensional process would allow direct control over two musical parameters (e.g., pitch and duration). It is possible, however, to obtain various levels of control by, for example, calculating various orbits of a lower dimensional system in parallel or by using systems of more than one equation. As a rule of thumb, devising mappings that are too simplistic may strip a potentially rich orbit of its details, producing music that is dull and uninteresting. Conversely, a method that is too complex may mask the behaviour of the orbit and jeopardise the composer's original intention to use the iterative process in the first place. Clearly, a balance must be struck.

The choice of initial values for the iterative process is also an important matter for consideration. As we saw earlier, orbits whose initial values lie close together may vary drastically after a few iterations. Sometimes it may be possible that a bad musical outcome was not due to faulty systematisation, but due to a bad choice of initial values. This phenomenon, however, can be used to the composer's advantage. For instance, by varying the initial values of the same iterative musical system, the composer can produce variations in a passage, or even in the entire piece, which begins in a similar manner and then diverges.

4.2 Fractal geometry

In the opening of the book *The Fractal Geometry of Nature*, Benoit Mandelbrot (1982) explains that a shortcoming of Euclidean geometry is its inability to describe the intricate shape of most natural things such as clouds, mountains, coastlines or trees. To address this inadequacy, Mandelbrot conceived and fostered the development of a new geometry that could describe many of the irregular and fragmented patterns of nature: *fractal geometry*.

A good way to become initiated in fractal geometry is to study one of its typical properties, namely *self-similarity*. Fractals carry fine patterns that repeat at different levels and sizes. Essentially, a fractal roughly resembles Russian dolls in the sense that fractals contain nested similar patterns comparable to the way in which a large doll contains a smaller version of itself, which in turn contains an even smaller version, and so on.

One way of modelling self-similarity is by means of iterative functions, hence the close link between iterative processes and fractals mentioned earlier. There are basically three types of self-similarity: *exact self-similiarity*, *statistical self-similarity* and *generalised self-similarity*. In exact similarity, each scaling factor reproduces exactly the same type of shape, as in the Sierpinski gasket shown in Figure 4.8. In statistical similarity, the resemblance at different scales is not exact, but it is close enough to be recognised. In nature, statistical self-similarity is frequently found in geological formations, in some microscopic living beings, and in flora and plants in general; cauliflowers, for example, have branches in forms that are very similar to the whole. Generalised self-similarity occurs when scaled copies of the whole object or figure undergo some transformation. Generalised self-similarity is more difficult to identify without knowing the transformations that might have taken place, but nevertheless there is often a generalised or categorical feeling for similarity when we perceive these objects.

Another characteristic of fractals is that they have fractional dimensions. The notion of a fractional dimension can be difficult to come to terms with, because we are used to thinking in terms of whole dimensions and normally not more than three. An anedoctal example to grasp the idea of a fractal dimension can be given as follows: imagine the task of measuring the perimeter of an island. In order to do this one has to establish the scale of the measurement. Zooming in on the details of the coastline results in an increase of the complexity and the length of the border line: should the border be the line coasting the rocks or should it be the line coasting grains of sand? As the unit becomes smaller, the perimeter increases astronomically. Intuitively, the coastline seems to be a one-dimensional line, but as the process of decreasing the unit continues indefinitely, fractal theory teaches us that the coastline gets increasingly close to filling a two-dimensional region. Hence the rationale that a fractal lies somewhere between one and two dimensions.

4.2.1 The Sierpinski gasket

A well-known example of a fractal is given by Sierpinski's rules for generating a sequence of self-similar triangular structures called Sierpinski gaskets (Figure 4.8). The sequence starts with a solid black triangle. Then, this triangle is split into four smaller triangles but the middle one is removed, as shown in the second image of the sequence in Figure 4.8. The same procedure is applied again to each of the three remaining triangles to form the third image of the sequence. This same procedure is then applied to each of the smaller black triangles, and so on. The more this process is repeated, the more detailed the image becomes. This process could be applied indefinitely to create greater and greater detail so that triangular structures within triangular structures emerge *ad infinitum* when we magnify the image. In Figure 4.8 the triangular structures are represented in two dimensions, but mathematicians calculate that the dimension of the Sierpinksi gasket is 1.584. There are a number of methods for calculating fractal dimensions; more information on this and fractals in general is well expressed in Barnsley's book entitled *Fractals Everywhere* (1988).

Figure 4.8 The Sierpinski gasket is composed of infinite self-similar triangular structures.

The initial triangle in Figure 4.8 is broken into three congruent figures, each of which is exactly one half the size of the triangle and one quarter of its area. By doubling the size of any of the three new figures we obtain an exact replica of the original triangle. If we regard the Sierpinski gasket as ternary and binary units reiterating on diminishing scales, we can relate it to musical form. A Sierpinski-like musical form could be given as an ABA composition, with repeated ternary and binary subdivisions. A discussion on deriving music from fractal forms is given later.

4.2.2 The Mandelbrot set

The Mandelbrot set, devised by Benoit Mandelbrot, is another typical example of a fractal structure (Figure 4.9). As mentioned earlier, a Mandelbrot set is defined by the following iterative equation: $z_{i+1} = z_i^2 + c$, where c is a complex constant. In practice, for each value of c the equation is initialised with $z_0 = 0$ and iterated. The process works as follows: applying the function $z^2 + c$ to $z_0 = 0$ yields the new number $z_1 = z_0^2 + c$, which in turn is used as the input for the next iteration $z_2 = z_1^2 + c$, and then $z_3 = z_2^2 + c$, and so forth. Depending on the values of c, the iterations generate either values that increase indefinitely or values that remain bounded. For example if $c = 1$ then:

$$z_0 = 0$$
$$z_1 = 0^2 + 1 = 1$$
$$z_2 = 1^2 + 1 = 2$$
$$z_3 = 2^2 + 1 = 5$$
$$z_4 = 5^2 + 1 = 26$$
$$z_5 = 26^2 + 1 = 677$$
etc.

In the above example we can clearly observe that $c = 1$ will generate values that increase indefinitely. In this case we say that $c = 1$ drives the orbit towards infinity. The graphical plot for this example can be seen in Figure 4.7; the plottings of cases where the orbit remains bounded can be seen in Figure 4.6. One thing that the example above does not reveal is that complex numbers themselves are two-dimensional. A complex number is a pair of real numbers (x, y), that is, a point in the plane, also thought of as $x + yi$, with $i^2 = -1$ for the purpose of calculation; this will be clarified below.

The Mandelbrot set consists of those values of c for which the corresponding orbit remains bounded. In order to implement this on a computer, a maximum threshold value is used to represent infinity and a maximum number of iterations M is determined. In this case, the iteration process is halted when either z_n overtakes the threshold or the number of iterations reaches M. If by the time the number of iterations has reached M the value of z_M has not overtaken the infinity threshold, then the the value of c belongs to the Mandelbrot set.

The two-dimensional image in Figure 4.9 is obtained by considering c to be a complex number, whereby the real part is associated to the horizontal co-ordinate and the imaginary part to the vertical co-ordinate; hence the reason c is referred to as the complex constant of the equation. A complex number contains two components x and y, where y is multiplied by an imaginary number i. Thus, a point of co-ordinates x and y in a complex plane is represented by the complex number $z = x + yi$. An imaginary number is given as the square root of -1; this is imaginary because such a real number cannot actually exist: the square of any ordinary number cannot be negative. Unless the reader is willing to implement the Mandelbrot algorithm from scratch, it is not really important to understand complex numbers. It suffices to say that each value of c is assigned a colour according to the number of iterations required to send the point to infinity; black is assigned if a value remains bounded.

Figure 4.9 A typical display of the Mandelbrot set.

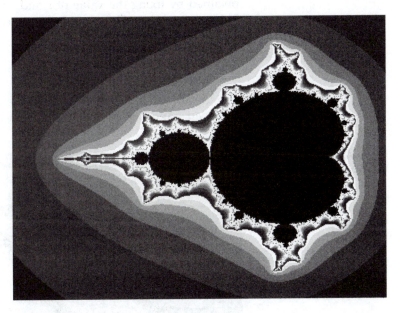

The Mandelbrot set corresponds to the black area of the figure. At the boundary of the Mandelbrot set there is a structure of enormous complexity. By magnifying sections at the edges of the

Figure 4.10 View of a magnified portion of the Mandelbrot set.

set boundary one can reveal a universe of infinitely embedded fractals (Figure 4.10).

A variant of the Mandelbrot set is the Julia set. The Julia set is obtained by fixing the value of c and varying the initial seed z_0 for each iteration. Figure 4.11 illustrates a Julia set where $c = -1.037 + 0.17i$.

4.2.3 Creating fractal musical forms

The idea of an iterative function whose results fall into classes, or dwell bands, is appealing, but the plot thickens when it

Figure 4.11 An example of the Julia set.

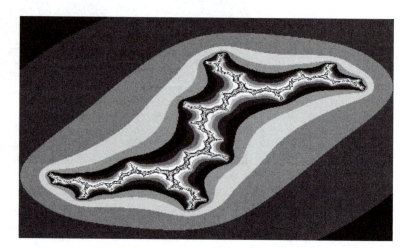

comes to the effective application of this idea for composing music. A simple way of exploiting the Mandelbrot equation for generating musical material would be to iterate the equation and associate the number of iterations required to make the point reach the infinity threshold with a particular pitch or class of pitches. This method, however, will hardly generate interesting music. One of the main caveats of using fractals in music is that the beauty that is perceived in their visual representation is given by the fact that the whole figure can be seen at once. Music is, par excellence, a time-based art form and as such it requires completely different perceptual strategies than a static figure requires for appreciation. Listeners would hardly hear any of the visual effects they can see in fractals if composers simply associated notes to the colours of the pixels of a fractal figure. A better approach to composing music with fractals is perhaps to by-pass the misleading beauty of the visual representation and seek inspiration in the fundamental mechanisms that generate self-similarity and scaling invariance. A few ideas inspired by the music of the past centuries are discussed below.

Self-similarity and scaling invariance were used to some extent in the music of previous centuries. One famous example of this is the chorale of Johann Sebastian Bach that appears at the end of his Art of Fugue, composed in the mid-eighteenth century. The chorale melody is slightly ornamented in the upper voice. The other voices prepare for the entry of the chorale by imitating the melody twice as fast. The alto voice plays the inversion of the melody and most of the other accompanying material is derived directly from the opening measure. The entire piece consists of three more phrases, all similarly arranged (refer to excerpt in Appendix I). In this case, self-similarity emerges from the repeated use of the same motifs within one larger section of the piece. The self-similarity of this piece is, however, only partial. It concerns the motif material of the piece, but not its harmony or its overall form.

Another reasonable route to fractal music is to look at the iterative properties of their functions and try to generate musical material by means of similar mechanisms. Fractal-like forms of musical interest may be generated by, for example, applying an iterative rule to an initial motif, or seed. This kind of generative process is described in mathematics as the translation of a basic geometrical shape with symmetry-group operations, such as *translation*, *reflection* and *rotation*, largely used in computer graphics. These transformational operations can be mathematically defined as given in Table 4.1.

Table 4.1

Reflexion around the y axis	$(x, y) \Rightarrow (-x, y)$
Reflexion around the x axis	$(x, y) \Rightarrow (x, -y)$
Rotation of 90°	$(x, y) \Rightarrow (y, -x)$
Rotation of 180°	$(x, y) \Rightarrow (-x, -y)$
Rotation of 270°	$(x, y) \Rightarrow (-y, x)$
Horizontal Translation	$(x, y) \Rightarrow (x + a, y)$
Vertical Translation	$(x, y) \Rightarrow (x, y + a)$

Analogous operations are widely used in music composition: transposition, inversion, retrograde, retrograde-inversion, augmentation and diminution; some of these were introduced in Chapter 2. Such operations on an initial motif are commonly found in most contrapuntal music and imitation is a ubiquitous operation in canons and fugues. As a matter of fact, from the Renaissance period through to the time of Bach, composers used to design enigma canons, or puzzle canons, in which a simple musical figure was given as a seed to performers who had to know the rules for making canons in order to figure out a solution to develop the seed by themselves; they had to decide on the intervals for the imitations and whether and when to apply various symmetry operations. Bach himself had an incredible talent for improvising on such seeds. As an example, Figure 4.12 portrays a puzzle canon whose authorship is attibuted to Johann Sebastian Bach. This was proposed as the seed for an eight-voice canon: it contains only pitch and metric information. Figure 4.13 presents a possible solution for the puzzle, provided by composer Larry Solomon (University of Arizona), where the time interval for the voice entrances is one quarter-note. In this case, the pitch interval for imitation is a perfect fifth, cycling back to tonic and the imitation is then inverted and transposed on alternating entries.

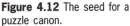

Figure 4.12 The seed for a puzzle canon.

Rhythm is undoubtedly an interesting domain on its own for composing with fractal-like structures. In 1941 Joseph Schillinger published the *Schillinger System of Musical Composition*, a 12-volume work that was years ahead of its time. Schillinger's book presaged many developments of algorithmic composition which would not be expanded upon until the computer became available. In the first volume, Schillinger dealt primarily with the theory of rhythm. Schillinger had the idea that composite

Figure 4.13 A possible solution for the puzzle canon of Figure 4.12.

rhythms could be formed by superimposing two different streams and forming a new stream that combined the timing onsets of both. By way of an example, let us consider a stream of period three and a stream of period four. If we consider 12 as the lowest common denominator for these two streams, the first one could consist of three notes of length four and the second of four notes of length three. Next, if we merge these two streams, the resultant stream will be composed of a note of length three, followed by a note of length one, then two notes

Figure 4.14 Composing rhythm streams.

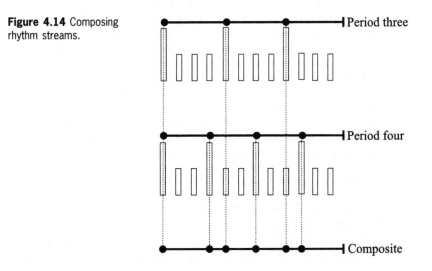

of length two, another note of length one and a final note of length three (Figure 4.14).

Another suggestion given by Schillinger was to construct rhythmic patterns by means of distributive powers. As an example of this, consider a series of fractions whose sum is equal to one. A new pattern can be produced by squaring this series as follows: since $(a + b) = 1$, then $(a + b)^2 = a^2 + ab + ab + b^2 = 1$. For example, starting with a 2–4 pattern, we form the binomial

$$\left(\frac{2}{3} + \frac{4}{3} \right)$$

By squaring this binomial we get

$$\left(\frac{2}{3} + \frac{4}{3} \right)^2 = \frac{4}{9} + \frac{8}{9} + \frac{8}{9} + \frac{16}{9},$$

which gives rise to a new rhythmic pattern: 4-8-8-16. In fact, any polynomial of any order adding up to one can be squared and cubed in order to generate new rhythmic patterns. The squares of $\frac{1}{4} + \frac{2}{4} + \frac{1}{4}$ seem to have been intuitively used by composers very frequently in the past. There is certainly a strong case for a multi-levelled fractal-like approach to musical structure in Schillinger's work which has been largely overlooked by computer musicians.

4.3 Related software on the accompanying CD-ROM

There are two excellent programs on the CD-ROM for generating music using iterative algorithms: FractMus (in the folder **fractmus**) and a Music Generator (in the folder **musigen**). Both programs are well documented and many examples are given. Also, there is MusiNum (in folder **musinum**) which is an interesting tool for exploring the potential of self-similarity in number sequences.

FractMus provides a number of iterative algorithms, mostly fractals, that can be selected to control the generation of a sequence. Many sequences can be generated at the same time, each of which can be controlled by a different algorithm. The composer can edit the parameters for the selected algorithms and these are well explained in the program's Help menu.

A variety of iterative algorithms is also available for generating music in a Music Generator, which is slightly more flexible than FractMus in the sense that users can define their own new functions and be more precise as to which aspect of the music will be controlled by the fractals. All algorithms can be fully customised and the program is praised for its ability to produce astonishing fractal pictures for preview.

5 Neural computation and music

5.1 Thinking music aloud

From a number of plausible definitions for music, the one that frequently stands out in the computer music arena is the notion that music is an intellectual activity, whose prominent tasks involve the abilities to recognise auditory patterns and to imagine them modified by actions. These abilities require sophisticated memory mechanisms, involving both conscious manipulations of concepts and subconscious access to millions of networked neurological bonds. In this case, it is true to say that emotional reactions to music arise from some form of intellectual activity.

Different parts of our brain do different things in response to sonic stimuli. Moreover, music is not detected by our ears alone; for example, music is also sensed through the skin of our entire body (Storr 1993) (Despins 1996). The response of our brain to external stimuli, including sound, can be measured by the activity of the neurones. The electrochemical behaviour of masses of neurones acting in small to very large groups in the brain produces a complex, multidimensional, pulsating electromagnetic field. Two methods are commonly used to measure this: PET (positron emission tomography) and EEG (electroencephalogram). Whilst PET measures the brain's activity by scanning the flow of radioactive material previously injected into the subject's bloodstream, EEG uses tiny electrodes (small metallic discs) pasted onto the surface of the skull by means of

Figure 5.1 The electroencephalogram gives a graph of the electrical activity of the areas of the brain where the electrodes have been placed.

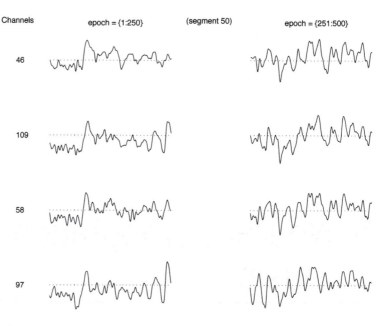

electricity conducting gel. As the signals captured by the electrodes are in the order of just a few micro volts, an amplifier is used to increase their amplitudes several hundred times. Whilst PET gives a clear cross-sectional indication of the area of the brain where the bloodflow is more intense during the hearing process, EEG gives a voltage level vs. time graph of the electrical activity of the areas of the brain where the electrodes have been placed (Figure 5.1).

Our understanding of the behaviour of the brain when we engage in any type of musical activity (e.g., playing an instrument or simply imagining a melody) is merely the tip of the iceberg. Both measuring methods have brought to light important issues that have helped researchers get this far. PET scans have shown that listening to music and imagining listening to music activate different parts of the brain and EEG graphs, combined with an analysis technique called ERP (Event-Related Potential), have been particularly useful in demonstrating that the brain expects to detect sequences of stimuli that conform to previously established experiences. As an anecdotal example to illustrate this, imagine that if you hear the sentence 'A musician composes the music', then the electrical activity displayed by the EEG/ERP maps would run fairly steadily. But if you hear the sentence 'A musician composes the dog', the activity of your brain would probably display a significant negative electrical response immediately after the word 'dog'. The human brain seems to respond similarly to musical incongruities. Such behaviour obviously depends upon our understanding of the overall meaning of the musical style in question. A number of enthusiasts believe that we are born programmed to be musical, in the sense that almost no-one has difficulties in finding coherence in simple tonal melodies. This is, however, a rather provocative and controversial statement which awaits stronger scientific validation.

One of the greatest scientific challenges of the twenty-first century is to gain a better understanding of the functioning of the brain when engaging in musical activities. Imagine, for example, if you could make music by means of a sort of mental instrument connected to your brain; no hands required. Actually, some systems of this nature already have begun to appear over the last decade (e.g., the IBVA system, discussed in Chapter 8) and research into the development of increasingly sophisticated brain analysis technology is on the agenda of research centres worldwide. The development of this technology brings a great number of benefits such as adding to the understanding of music and the mind, and promoting the development of thought-controlled systems for aiding those with physical difficulties. Many examples in both musical and non-musical research fields have confirmed that EEG measurements provide a rich source of information about our thought processes: principal components analysis of target detection in auditory stimulus experiments (Jung *et al* 1997), statistical analysis and brain mapping of information flow between EEG channels during music listening experiments (Saiwaki 1997), and the analysis of time averaged EEG's from listening experiments (Janata 1995), to name but three.

Brain signals have been tentatively categorised into four main types:

1 a random-seeming background,
2 long-term coherent waves,
3 short-term transient waves, and
4 complex ongoing waves components (Rosenboom 1990).

The random-seeming background, about which little is known, is the residue observed after all known methods of waveform decomposition have been exhausted. Long-term coherent waves are the well-known *alpha, beta, delta,* and *theta* rhythms, which range from approximately 1 Hz to approximately 40 Hz. They are often associated with certain states of consciousness, such as alertness and sleep. Short-term transient waves reflect the specific experiences associated with external stimuli. Until now these waves have only been accessible by means of EEG/ERP maps. ERP analyses are achieved by taking the average of many EEG recordings, where the person is repeatedly subjected to the same stimulus. The reason for averaging is to remove noise caused by other non-correlated brainwave activities. ERP analysis has played an important part in studying music cognition, as authors like Rosenboom (1990) and others have shown. Finally, it is suggested that a non-random complex component exists, whose ever-changing pattern comes from the build up of baseline activation of the vast neuronal masses within the brain; little is known about this component.

Although powerful techniques for recording brainwaves exist, in practice it is difficult to infer how our brain represents the music we hear or imagine. For a start, the brain never works exclusively on processing music. Given the complexity of the EEG signal, finding recurring patterns, let alone linking specific patterns with musical thought, is an incredibly difficult task. Broadly speaking, brain/computer interfaces can be divided into three operational categories:

1 those where the computer adapts to the user,
2 those where the user adapts to the computer and
3 those where both the computer and the user adapt to each other.

People who have attempted to utilise the EEG as part of a music controller have done so using biofeedback systems that map certain EEG signals such as the alpha wave (approximately 10 Hz) to specific musical actions, allowing the user to experience the results of their performance in real-time. In this case, controlling the music requires the user to learn to direct their EEG to produce certain characteristic patterns: the second category. The IBVA system on the accompanying CD-ROM is an example of such a system (Chapter 8).

Another facet of neurology-based computer music comprises the simulation of the brain architecture on computers by means of artificial neural networks.

5.2 Artificial neural networks

From the discussions in the previous chapters one can conclude that the representation of explicit and unambiguous data, rules and/or procedures is an imperative practice that is normally taken for granted by software designers. The classic approach to the design of a musical composition program often boils down to putting together a set of subroutines that embodies a system of compositional rules, which in turn are called upon by generative engines that produce the music. This engine normally functions by generating random musical possibilities that are evaluated against the rules: those possibilities that do not violate the rules are incorporated and the offensive ones discarded. One can aggregate a number of sophisticated mechanisms to this basic idea, such as learning algorithms that automatically build the compositional rules by looking at given examples. This approach to software design is normally called rule-based or knowledge-based.

The main criticism of the rule-based approach to composition software is that the formalisation of music often implies reductive processes that fail to capture numerous informal aspects of the musical phenomena. Quoting Loy, from the book *Music and Connectionism* (1991, pp. 28) the formalisation of certain musical domains 'limits the usefulness of reduced models to trivialisation of complex musical problems'. Basically, Loy is suggesting that although rules could be taken as the foundation of a piece of music, art is not merely about following rules. One must consider that music does not exist in a vacuum, but is based on the consensual experience of a whole culture, which is not always explicit.

Neural networks, also known as connectionist systems, constitute an alternative approach to the traditional theory of computation and indeed to the rule-based approach to algorithmic composition. The main claims in favour of neural networks are that they 'eliminate the Central Processing Unit vs. memory dichotomy and the concept of symbolic instruction set', and that they 'provide learning mechanisms so that the desired computation can be programmed by simply repeatedly exposing the network to examples of the desired behaviour' (Swingler 1996). In principle, connectionist systems are not constructed to learn by formulating collections of explicit rules. Rather the result of the learning should be such that it appears in hindsight, as though the model did indeed know those rules.

Although the author can hardly agree that artificial neural networks, as we know them today, can solve the problem of the trivialisation cited above, they do, however, have the potential to outperform the rules-based approach in certain tasks. It is sensible then to regard the connectionism approach as complementary to the rule-based approach, instead of as a replacement.

5.2.1 Understanding the brain

This section presents a brief introduction to the functioning of the human brain with the intention of giving an insight into the inspiration for the design of artificial neural networks. Note, however, that artificial neural network systems normally omit many aspects of the real physiological brain for various reasons; for instance, the notions of pulse streams and refractory periods (discussed below) are seldom included in these systems.

The brain is formed by a massive and dense network of interconnected neurones. Each neurone can be considered as an independent processing unit on its own and there are various different types of neurones in our brain system. Connections between the neurones are established by *synapse junctions* or simply, *synapses*. A neurone receives stimulation either from the outside world or from other neurones through the intricate branching of neurofilaments extending along both its body (also referred to as *soma*) and dendrites, establishing the synapses (Figure 5.2). The stimulation is in the form of electrical pulse streams at variable rates, and a neurone discharges electrical pulses to other neurones via a single slender fibre, the *axon*, which makes contact with numerous receiving neurones.

Figure 5.2 When an electrical impulse arrives at the tip of the neurone's axon, it triggers the discharge of neurotransmitter molecules which stimulate the dendrites of the receiving neurone.

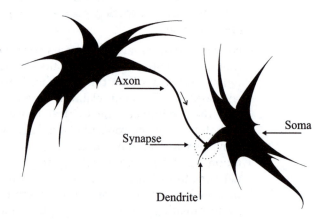

Coming back to the notion that neurones are electrically-active cells, we are now in a position to study how they interact with one another through the flow of local electric current. These electric currents are driven by a voltage difference across the synaptic membranes at the synaptic junctions (Figure 5.3). The axon from a neurone does not actually touch the body of the receiving neurones; there is a fluid-filled cleft separating them. As stimulation reaches the end of an axon, a chemical substance, the *neurotransmitter*, is released and diffused across the synaptic cleft. This induces a voltage change across the synaptic membrane of the receiving neurone. However, the internal voltage of this neurone is normally maintained at a certain level, referred to as *equilibrium potential*. This equilibrium potential is determined by the action of certain biochemical components that exchange electrical ions across its membrane. The neurones have control over the ionic permeability of their synaptic membranes in order to maintain their equilibrium potential. Basically what one neurone 'wants to achieve' by sending stimulation to another is to disturb the equilibrium potential of the latter. If the current of the stimulation is sufficiently high to achieve this, then the disturbed neurone will try to restore its equilibrium by discharging stimulation through its own axon, which in turn will probably disturb the equilibrium of others. The whole brain activity resembles the effect of infinitely vast streams of falling dominoes, but in this case the falling dominoes stand up again after a short lapse of time, then fall again, and so on.

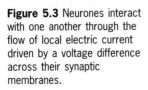

Figure 5.3 Neurones interact with one another through the flow of local electric current driven by a voltage difference across their synaptic membranes.

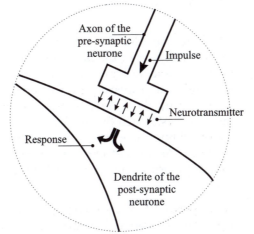

Axon of the pre-synaptic neurone

Impulse

Neurotransmitter

Response

Dendrite of the post-synaptic neurone

A nerve pulse thus results from a rapid voltage change produced by the flow of electric current through the synapses, caused by an internal voltage disturbance. In essence, a neurone is an all-or-nothing unit. It is said to *fire* (that is, to emit stimulation down its axon) if its body is sufficiently depolarised by the arriving stimulation. Depolarised here means becoming increasingly charged with electricity. It should be noted, however, that the incoming pulse streams can be either *excitatory* (causing depolarisation) or *inhibitory*. Under a barrage of incoming stimuli, a neurone adds the incoming electrical currents arriving from all over the place; the neurone then sums all incoming excitatory signals and subtracts all the inhibitory ones. If the result exceeds its tolerance to electrical disturbance then the neurone fires; that is, it emits stimulation to others. The tolerance of a neurone to electrical disturbance is referred to as the *firing threshold*.

Neurones combine the pulse streams that reach them using different mechanisms (Carpenter 1990). For example, the *temporal summation mechanism* (TSM) combines the frequencies of the incoming stimulation. Once a single pulse arrives at its destination it tends to fade away. But if the next pulse arrives before the preceding one has completely faded, then the two voltages will add up. A pulse stream of sufficiently high frequency should then build up energy at the edge of the recipient neurone and this neurone will eventually fire. The time interval between the synaptic activation of a neurone and the activation of the next neurone in sequence may be no more than a millisecond, but after firing, a neurone will resist activation for several milliseconds. This period is called the *refractory period*. During this period the neurone restores its equilibrium potential. The alternate states of refraction and excitability are probably reflected in the brain as a whole in the rhythm of the brain waves.

The brain can thus be studied as a massive array of parallel processing devices: the neurones. Curiously, neurones typically operate several orders of magnitude slower than the silicon logic gates used in ordinary personal computers. But this is compensated for by the utterly enormous inter-connectivity between them. It is estimated that the human cortex contains in the region of 10 billion neurones interconnected by about 60 trillion synapses. As a result humans can perform certain tasks such as sound pattern recognition much more efficiently than computers.

An important biological concept to bear in mind when designing artificial neural networks is the concept of *ganglia*. The biological nervous systems of all but the most primitive organisms are modular in structure. That is, they consist of a set of interconnected sub-sets, each performing a different task. In complex organisms, groups of functionally-related neurones are clumped into ganglia. Ganglia are connected to one another through nerves, which act as communication channels. In highly advanced organisms there is one overgrown super-ganglion: the *brain*. The brain therefore is not uniform but consists of functional centres linked by nerve trunks, each one responsible for one or more specific tasks.

5.2.2 Building artificial neural networks

Any attempt to model the brain should start with a satisfactory representation of the neurone. One of the first artificial neurone models appeared in the 1950s and it was called the *perceptron*

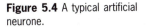

Figure 5.4 A typical artificial neurone.

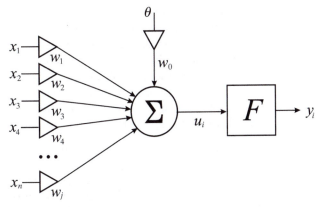

(Rosenblatt, 1958). Since then a number of variations and improvements on the basic perceptron have been proposed. Readers are invited to refer to the tutorial articles by Lippmann (1987) and Hush and Horne (1993) for a comprehensive overview. A typical neurone model is illustrated in Figure 5.4.

The synapses are represented by the connecting patches. Note that each patch has a weight value associated with it, which scales the signal arriving at the input. This weight represents the strength of the incoming stimulation. A signal x_n at the input of the jth synapse of neurone i is multiplied by the synaptic weight, w_{ij}. This weight may be either positive, if the synapse is *excitatory*, or negative, if the synapse is *inhibitory*. In other words, weight may either boost the input signal or attenuate it. In addition to the signal inputs there might also be a bias term θ, which has the effect of lowering or raising the net input to the activation function. One could loosely associate this bias to the resistance of the neurone to incoming stimulation. All weighted input signals are then added at the *summing junction* (Figure 5.4). The result of the summation u_i is then passed through an activation function (also called transfer function) F, whose objective is to normalise the output of the neurone to some prescribed range of values, typically between 0 and 1 or between –1 and 1.

The neurone in Figure 5.4 is mathematically written as follows:

$$u_i = \sum_{n=0}^{j} w_{in} x_n$$

which corresponds to the summing junction:

$$y_i = F(u_i)$$

and which corresponds to the activation function.

The weight w_{ij} of a synapse may have either a fixed or dynamic value. Dynamic synapses are used to establish a learning mechanism. For example, one could increase the weight between synapses whose activation levels change in a temporally-correlated manner, tending to subsequently couple the activity of the neurones involved. Learning mechanisms may also be implemented by altering other features of the unit, such as the threshold and the temporal summation mechanism.

The activation function F defines the behaviour of the neurone in terms of its input. This corresponds to the tolerance of the neurone to electrical disturbance, as mentioned earlier: the *firing threshold*. There are basically three types of activation functions: threshold, *piece-wise linear* and *sigmoid*.

A threshold function only passes on information if the incoming activity level reaches a threshold:

$$F(x) = \begin{cases} 0 \leftarrow x \leq 0 \\ 1 \leftarrow x > 0 \end{cases}$$

Here, $F(x)$ will value 0 if x is lower than or equal to 0, otherwise it will equal 1 (Figure 5.5).

Figure 5.5 Threshold function.

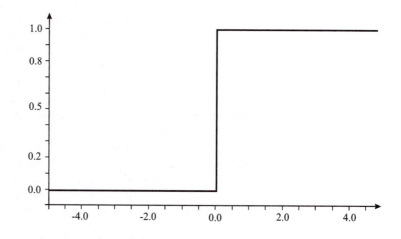

Piece-wise linear functions are defined as follows:

$$F(x) = \begin{cases} 0 \leftarrow x \leq 0 \\ x \leftarrow 0 < x < 1 \\ 1 \leftarrow x \geq 1 \end{cases}$$

In this case, $F(x)$ will value x if x is a value between 0 and 1; 0 if x is lower than or equal to 0; or 1 if x is equal to or higher than 1 (Figure 5.6). Note that a piece-wise linear function may have more 'pieces'.

Figure 5.6 The piece-wise linear function.

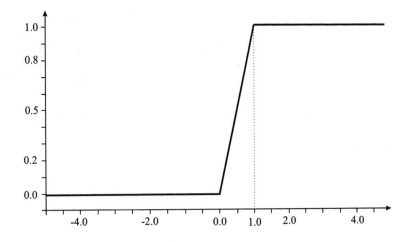

Sigmoid activation functions are smooth, increasing functions with horizontal asymptotes. Sigmoids are the most widely-used activation function (Figure 5.7). Example: $F(x) = (1 + e^{\delta x})^{-1}$; this function varies from 0.0 to 1.0 according to the value of x. The coefficient δ determines the steepness of the transition area: as this value increases, the sigmoid curve approaches non-linearity (typical value is $\delta = 1.0$).

Figure 5.7 An example of a sigmoid function.

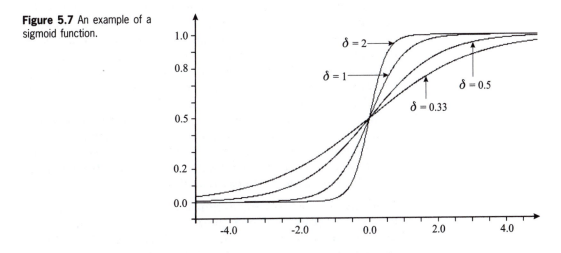

The output of a neurone is then connected to the inputs of other neurones. The network design, that is the manner in which the neurones are interconnected, defines the behaviour of the network. Moreover, there is an infinite number of possible variations on the basic neurone introduced above. Various applications call for different designs according to the task at hand.

Most artificial neural networks are designed as a type of automaton and, as such, they can be represented as weighted graphs (Chapter 2), where each node encapsulates an individual neurone and the weighted edges represent the synaptic links. In this case, there is also an activation function associated to each node in order to scale the output values. Figure 5.8 portrays an example of a simple neural network represented as a graph.

Figure 5.8 The graph representation of a 2–2–1 multi-layer feed-forward network.

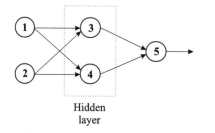
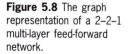

Note that in Figure 5.8 there are three layers: one input layer, one hidden layer and one output layer. The nodes of the input layers are the eyes and ears of the network, whereas the output ones give us the results. Between the input and the output layers there are usually one or more hidden layers made of nodes that have no external connections. The nodes of a given layer of the network are connected to the nodes in higher layers in a forward direction. In this case, information flows in only one direction through the network; there are no feed-back loops. This characterises the network as *feed-forward*. Another significant point to observe is that the network is *fully connected* (as opposed to *partially connected*) in the sense that each node in one layer is connected to each node in the next layer. As a matter of convention, this is a multi-layer 2–2–1 network: the numbers refer to the quantity of nodes in each layer. The middle layer of nodes is referred to as a hidden layer because their behaviour is normally not visible to an external observer.

There are also *feed-back* architectures where the output of a node is connected back to its own input, either directly or indirectly. The output from the nodes of a feed-back network is always dependent on the previous state of the network (Figure 5.9).

Figure 5.9 An example of a feed-back architecture with direct feed-back connections.

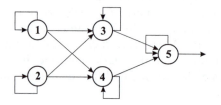

The examples given above are known as the Multi-Layer Perceptron (MLP) model. In addition to the MLP model, the Kohonen model is also widely employed for musical tasks (Toiviainen 2000).

Kohonen networks are ideal for tasks involving the classification of data into clusters of information containing similar features on a topographical map. They were inspired by the fact that for senses such as hearing and vision, there is topographical mapping between the neurones in the cortex and the cells in the sensory organ. Imagine for example, that when adjacent cochlear cells in the ear are stimulated by sound vibrations, they produce responses in neurones that are close together in the part of the brain that is carrying out the auditory processing. The classic Kohonen network consists of only two layers (Figure 5.10): one serves as the input layer and the other constitutes the topographic space where the results of the classifications will be settled. The neurones of the second layers are competitive in the sense that they compete among themselves to classify the current input pattern: only one neurone can win. The winner and the neurones in its neighbourhood are rewarded by having their connection to the input neurone modified. A new pattern is then presented and this process is repeated a number of times. The effect of rewarding neighbours as well as the winner, results in the network being capable of producing topographical mappings. At the start of the learning process, a large neighbourhood is defined so that many of the winner's neighbours have their connections modified. As learning progresses, the neighbourhood size decreases so that increasingly fewer neighbours are rewarded. The final result is a topographical map where the neurones that are close together in the competitive layer respond to the patterns that are close together in the input space.

Figure 5.10 A Kohonen network with only two layers.

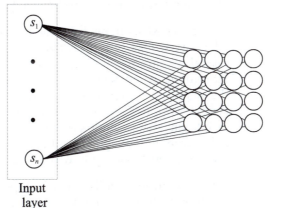

Input
layer

5.2.3 Training the network

In general terms, training a neural network involves presenting the network with a series of samples of the problem to be solved and an example of a solution for each sample problem. Given enough training material, the neural network should be able to learn the underlying aspects of the solutions. Should a similar problem appear, the network then retrieves these aspects to solve it.

In general, there are three basic stages in the learning process of a neural network:

Step 1: the network receives stimulus at its input neurones
Step 2: this input causes changes in the state of the network neurones (e.g., changes in the weights)
Step 3: the network responds differently due to the changes that have taken place

There are three fundamental approaches for implementing learning schemes using neural networks: *supervised learning, reinforcement learning* and *unsupervised learning*.

Supervised learning assumes that there is a teacher that constantly supervises the whole training process to ensure that the network behaves in the manner that is required. In this case the teacher must know in advance what the network is supposed to learn. *Reinforcement learning* also requires a supervisor, but in a less strict fashion. Here the supervisor only indicates whether the network has produced the expected behaviour from a given stimulus, thus reassuring the network when it produces the right outcome. And finally, as the name suggests, an *unsupervised learning* situation requires no help from external supervisors; the network is supposed to learn entirely by itself. We will only focus on supervised learning here, due to its simplicity and practicality as a basis for musical applications.

In supervised learning the training data consists of examples which have a set of input values and an associated output. The typical architecture for supervised learning is the MLP model introduced in the Section 5.2.2. As the inputs are presented to the network it produces an output; this is compared to the correct output. The teacher then causes the network to change its internal representation of the data so as to capture the essential features of the input data. In practice, there are usually learning rules which specify how the weights should be updated.

A typical musical application of a supervised MLP is to train the network with a set of musical examples in order to produce similar musical material. In this case, the network must optimise its outcome in response to the training data. This optimisation is accomplished thanks to a mechanism called *error-correction*. An error signal $e_k(n)$ can be expressed as follows:

$$e_k(n) = d_k(n) - y_k(n)$$

where $d_k(n)$ denotes the desired response from a certain neurone k at time n, and $y_k(n)$ denotes the current response of this neurone at time n. The ultimate purpose of the error-correction mechanism is to minimise a *cost function* based on the error signal such that the response of each output neurone in the network approximates the desired overall response in some statistical manner. The cost function that is normally employed in this case uses the mean-square value of the sum of the square errors, defined as follows:

$$J = E\left[\frac{1}{2}\sum_k e_k^2(n)\right]$$

where E denotes the expectation and the summation Σ is over all neurones in the output layer of the network. The network is optimised by minimising J with respect to the synaptic weights of the network. Thus, according to the error-correction rules, an adjustment Δ should be made to the synaptic weight w_{kj} at time n as follows:

$$\Delta w_{kj}(n) = \eta\, e_k(n)\, x_{kj}(n)$$

where η is a constant that denotes the rate of learning (this is normally specified by the teacher beforehand) and $x_{kj}(n)$ denotes the pre-synaptic activity arriving at the jth synaptic patch of neurone k at time n.

Considerable care must be taken when setting the value of the constant η for the rate of learning in order to ensure the stability of the learning process. If this constant is too small, then the learning process will proceed smoothly, but it may take a long time until the system actually reaches stable behaviour. Conversely, if η is too large, then the rate of learning will be accelerated, but this also increases the risk of causing the whole learning process to diverge, thus leading to unstable behaviour. The best way to proceed here is to try a few options and select the one that performs best.

5.3 Musical networks

As a practical example of an application of a neural network to generate music, let us study a hypothetical example where one wishes to train a network to produce music in a particular style by feeding the system with examples of compositions in the style that the system should imitate. This is illustrated below with a neural network proposed by Kenny McAlpine (1999) at the Department of Mathematics of Glasgow University, which is intended to generate simple monophonic melodies (Figure 5.11).

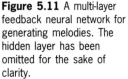

Figure 5.11 A multi-layer feedback neural network for generating melodies. The hidden layer has been omitted for the sake of clarity.

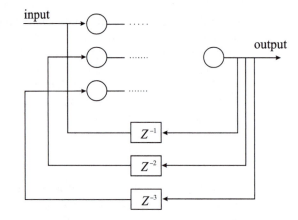

As the network is intended to produce only monophonic melodies, there is only one output neurone, whose outcome is taken to be the next note in the melody. This note is then fed back to the input layer for the computation of the following note, and so on. On its way back to the input layer, the outcome value passes through a delay unit (Z^{-n}) that holds back the value by n output calculations. Without the delay unit, the outcome value would be passed back to the input layer for the current calculation rather than for a subsequent one. By using three delay units, our neural network takes into account the three previous notes of the outcome sequence when calculating the next one. In order to keep the example simple, we only deal with the calculation of the pitch attribute of the note; there is no provision here for duration, intensity, etc. These other attributes could be easily added, however, by running other networks in parallel, one for each attribute, or simply by adding a further processing path in the above network.

Pitches are represented as MIDI note numbers, where 60 corresponds to the middle C (Figure 5.12).

At the training stage, the synaptic weights are initialised with random values. The teacher then feeds a number of melodies to the network note by note and adjusts the synaptic weights so that the output corresponds to the next pitch in the melody. In

Figure 5.12 Musical notes and their corresponding MIDI pitch value.

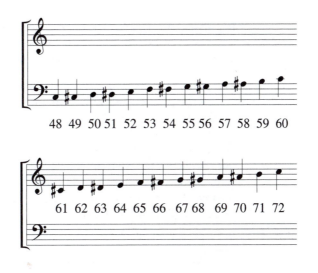

order to generate melodies, the neural network is given a starting note as input. The output from this note is then fed back to the input and used to calculate the next note and so on.

The following example, also proposed by McAlpine, is a neural network whose objective is to give a consonance measure to the interval between two given notes. This network could be used to filter undesired items produced by a random notes generator, for example.

Before we introduce the network, let us examine the function $\alpha(c_1, c_2)$ that McAlpine devised to measure the consonance of an interval between two notes c_1 and c_2:

$$\alpha(c_1, c_2) = A_1 \, P(|c_2 - c_1|(\text{mod } 12)) + A_2 \left[\frac{|c_2 - c_1|}{12} \right]$$

where A_1 and A_2 are constants, P is a preference coefficient, [ ] denotes 'integer part of' and $|c_2 - c_1|$ means the absolute value of the result of the subtraction. The basic interval is calculated by taking the module of the integer part of the subtraction of c_1 from c_2 ($|c_2 - c_1|(\text{mod } 12)$). The consonance measure $\alpha(c_1, c_2)$ is interpreted as a measure of the composer's subjective preference P for a specific interval ($|c_2 - c_1|(\text{mod } 12)$) plus a factor

$$A_2 \left[\frac{|c_2 - c_1|}{12} \right]$$

that gives the number of octaves over which the interval spans. This additional factor is necessary in order to take into account the different tonal qualities of intervals that span beyond the range of one octave. The constants A_1 and A_2 should be chosen experimentally, according to the preferences of the composer.

For instance, if $A_1 = 1$ then A_2 should be preferably set to the reciprocal of the number of octaves over which an interval may span before its harmonic character changes.

The values for the preference coefficients P are defined by fixing one of the pitches and letting the other pitch range through each of the twelve possibilities. Then, one assigns each pitch an integer between 0 and 11 to rank the intervals in order of preference. Note that these assignments do not have to be unique. If one decides that two intervals are equally pleasing, then the same rank can be assigned to both. An example can be given by setting P as follows (the lower the value of P the higher the preference for the interval):

Table 5.1

Interval:	Preference coefficient: P
0	4
1	8
2	6
3	4
4	6
5	4
6	6
7	5
8	4
9	5
10	6
11	8

Once the preference coefficients have been defined, one can use these coefficients to train a neural network to recognise the consonance measure for each of these intervals or any other interval that it may come across.

The neural network is given in Figure 5.13. The input pair c_1 and c_2 is fed into the two input nodes on the left. The nodes in the middle are labelled 0, 1, 2, ..., 11 from top to bottom. A node n will be active (i.e., equal to 1) if the input pair belongs to the corresponding interval class; e.g., if $|c_2 - c_1| = 2$, then node 2 will be active. All other nodes will be inactive (i.e., equal to 0). The next step is concerned with ordering the interval classes according to their preference coefficient P. This is achieved by scaling each of the output nodes n by the corresponding factor of $P(n)$. Then, the bias term is applied to the summation of the scaled nodes. Once trained, the network offers an automated device for calculating the consonance index of any given interval.

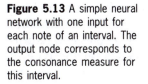

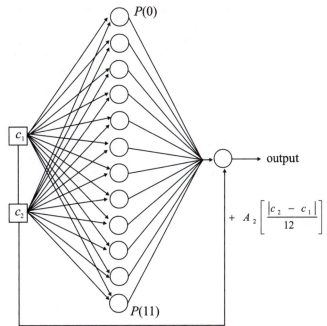

Figure 5.13 A simple neural network with one input for each note of an interval. The output node corresponds to the consonance measure for this interval.

This network is extremely simple and practically no real learning is taking place. Nevertheless, it is a good introductory example of the classificatory capabilities of neural networks. The real potential of this type of network appears once it has been expanded to cope with more than one interval; e.g. calculating the consonance of chords. Such a network could be used in chord substitution tasks for example, whereby it would examine all chords in the immediate neighbourhood of the pair in question and choose the pair that provided the best consonance coefficient.

5.4 Related software on the accompanying CD-ROM

A demonstration of the IBVA system for controlling music using brainwaves is given in the folder **brainwav**, in the Macintosh partition. Unfortunately there is no provision for doing this in practice because the electrodes are not provided. More information about the system can be found in the documentation.

Although much research has been done in the field of neural networks applied to music, no neural networks-based system for composition is readily available for a third party to use. There is certainly a case for someone to put together a package of this kind. In the folder **camus\disnet** there is a program for Windows called Dissonance Network, by Kenny McAlpine, which implements the neural network described earlier. Here two single notes are fed into the network, which then classifies the resulting interval formed between the two notes and returns a measure of the consonance, which depends on the type of interval formed and how physically far apart the notes are. Different classes of intervals are ranked according to the user's own preferences.

Readers wishing to learn more about neural networks applications for music composition are invited to study the following references: Baggi (1992), Bharucha and Olney (1989), Carpinteiro (1995), Dolson (1989), Freisleben (1992), Laine (1997), Linster (1990) and Todd (1989).

6 Evolutionary music: breaking new ground

Contemporary scientists seem to be increasingly shifting their attention from the study of the composition of matter to the study of the functional characteristics of nature's tangled systems, including the *interaction* between the components of a system, the *interconnection* of different systems and the emergence of *global behaviour*. Consequently, new scientific methodologies and tools are being devised for these new studies and most of them would not be possible without computer technology: scientists can now create surrogate artificial worlds in order to perform complex experiments that would otherwise be impossible. More enthusiastic academics often compare these surrogate systems to the biochemical laboratories used by scientists for centuries to investigate the structure of chemical components, cells, and so forth. The fundamental difference is that these systems are used to simulate the phenomena in question in order to study them in terms of the functional activity carried by patterns of information. The emergence of fields such as artificial life, or alife for short, is a consequence of this shift of scientific paradigm.

Alife is a discipline that studies natural living systems by simulating some of their biological aspects *in silico* (Langton, 1997). The attempt to mimic biological phenomena on computers is proving to be a viable route for a better theoretical understanding of living organisms, let alone the practical applications of biological principles for technology (robotics, nanotechnology, etc.) and medicine. Because alife deals with such complex phenomena, its development has fostered the development of a pool of research tools for studying complexity; for example, cellular automata, genetic algorithms, adaptive games and neural networks. It is interesting though, that these tools are also proving to be useful in fields other than biology, most notably social sciences (Gilbert and Troitzsch, 1999), linguistics (Steels, 1997) and musicology (Todd, 2000; Bilotta *et al.*, 2000).

Alife techniques may have varied applications in musicological research, but from a composer's point of view perhaps the most interesting application is for the study of the circumstances and mechanisms whereby music might originate and evolve in artificially created worlds inhabited by virtual communities of musicians and listeners. Origins and evolution are studied here in the context of the cultural conventions that may emerge under a number of constraints, for example psychological, physiological and ecological.

A better understanding of the fundamental mechanisms of musical origins and evolution is of great importance for composers looking for hitherto unexplored ways to create new music. As with the fields of acoustics (Campbell and Greated, 1987), psychoacoustics (Howard and Angus, 1996) and artificial intelligence (Miranda, 2000), which have greatly contributed to our understanding of music, alife has the potential to reveal new aspects that are waiting to be unveiled. Let us take as an example the origin and evolution of *musical styles*. Musical styles seem to result from the emergence of rules and/or from the shifting of existing conventions for *music making* (the term 'music making' is used in this chapter to mean both creating music and listening to music). Musical styles maintain their organisation within a cultural framework and yet they are highly dynamic; they are constantly evolving and adapting to new cultural situations. The hypothesis here is that the mechanisms for simulating the complexity commonly found in biological systems may also explain some aspects underlying the emergence of musical styles. One should be careful, however, not to overstate the potential of alife models here. It is unlikely that models entirely based upon biology alone would be able to realistically support any reasonable explanation for the origins and evolution of music. The great challenge for musicologists is to devise complementary approaches that can better address the problem at hand.

This chapter discusses the application of three alife paradigms for music: *cellular automata, genetic algorithms* and *adaptive games*.

Whilst the first two sections focus on cellular automata and genetic algorithms from a pragmatic generative point of view, the third section presents a more speculative discussion on the possibility of basing compositional processes on simulations of the fundamental origins of musical mechanisms and their evolution.

6.1 Cellular automata

Cellular automata (CA) are computer modelling tools widely used to model systems that change some feature with time. They are suitable for modelling dynamic systems in which space and time are discrete, and quantities take on a finite set of discrete values.

A cellular automaton consists of an array of elements, referred to as cells, to which evolution rules are applied. These rules determine the behaviour of the automaton in time. All the cells in the array are updated simultaneously, so that the state of the automaton as a whole advances in discrete time-steps. The fact that a CA cell can represent anything from a simple numerical variable to sophisticated processing units makes CA a very powerful modelling tool. In this section we will focus on the type of CA application where the cells stand for single variables that can hold an integer number. The class of CA studied here is often referred to as the p-state cellular automata because their cells can value a number p of possible integer values: 0, 1, 2, ... $p - 1$.

By way of an introductory example, Figure 6.1 illustrates a very simple CA: it consists of a bar (one-dimensional array) of 12 cells and each cell can value either zero or one, represented by the colours white or black, respectively.

From an initial random setting, at each tick of the clock, the values of all 12 cells change simultaneously from top to bottom on the computer screen, according to a set of rules that determines a new value for each cell. Cellular automata rules normally take into account the values of a cell's nearest neighbours, but these rules could also consider other cells, depending upon the kind of behaviour one is looking for. An example of a rule could be given as follows: 'if a cell is equal to zero and if both neighbours are equal to one, then this cell continues to equal zero in the next stage'.

The patterns formed by the black cells are the result of the automaton's *emergent behaviour* in the sense that the evolution rules in a CA are concerned only with local neighbourhoods

Figure 6.1 A simple one-dimensional CA. The right-hand figure is equivalent to the left-hand one, with the difference that it displays the colours associated to cell values, where 0=white and 1=black.

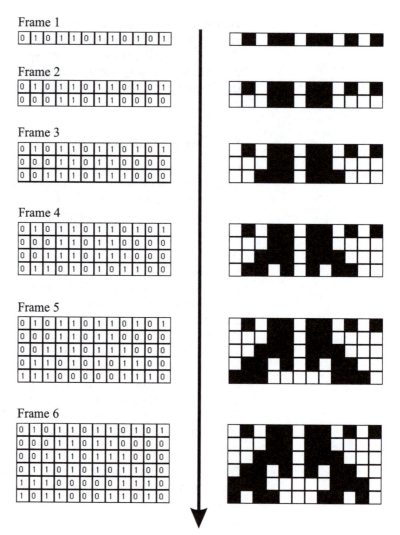

around the cell under consideration: no global trend is explicitly coded beforehand.

More sophisticated CA configurations use a two- or even three-dimensional array of cells that can assume values other than zero and one; in this case the values are represented by various different colours (Figure 6.3). In the case of a two-dimensional array, the evolution rules normally take into account four or eight neighbours (Figure 6.2).

The functioning of a cellular automaton is normally monitored on the computer screen as a sequence of changing patterns of tiny coloured cells, according to the tick of an imaginary clock,

Figure 6.2 An example of a two-dimensional array of cells where the evolution rules take four neighbours into account.

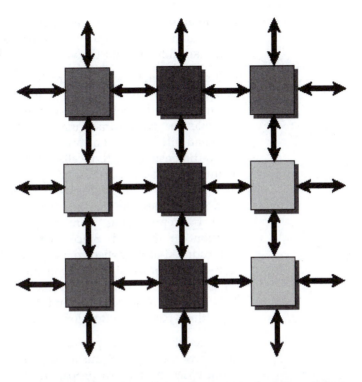

Figure 6.3 More sophisticated CA configurations hold cells that can assume values other than zero and one.

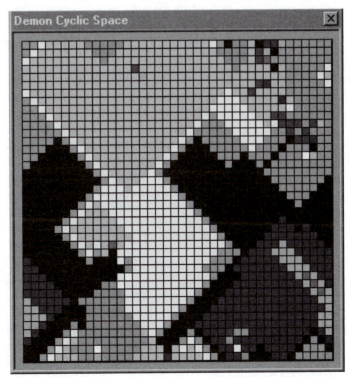

like an animated film. At each tick of the clock, the values of all cells change simultaneously, according to a set of transition rules that takes into account the values of their neighbourhood.

Cellular automata were originally introduced in the 1960s by von Neumann and Ulan as a model of a self-reproduction machine (Cood, 1968). They wanted to know if it would be possible for an abstract machine to reproduce; that is, to automatically construct a copy of itself. Their model consisted of a two-dimensional grid of cells, each cell of which could assume a number of states, representing the components from which they built the self-reproducing machine. Completely controlled by a set of rules, the machine was able to create several copies itself by reproducing identical patterns of cells at another location on the grid. Since then, CA have been repeatedly reintroduced and applied to a considerable variety of purposes, from image processing (Preston and Duff, 1974) and ecology (Hogeweg, 1988) to biology (Ermentrout and Edelstein-Keshet, 1993) and sociology (Epstein and Axtell, 1996). Many interesting algorithms have been developed during the past 30 years.

Since cellular automata produce large amounts of patterned data and if we assume that music composition can be thought of as being based on pattern propagation and the formal manipulation of its parameters, it comes as no surprise that researchers started to suspect that cellular automata could be associated to some sort of music representation in order to generate compositional material.

The remainder of this section will present the basic functioning of CAMUS, a cellular automata music generator originally designed by the author and further developed by Kenny McAlpine and Stuart Hoggar at the University of Glasgow. Of the many different cellular automata algorithms available today, two were selected for CAMUS, namely Game of Life, invented by John Horton Conway and Demon Cyclic Space, designed by David Griffeath (Dewdney, 1989).

6.1.1 Game of Life

The Game of Life is a two-dimensional CA that attempts to model a colony of simple virtual organisms. In theory, the automaton is defined on an infinite square lattice, but for practical purposes it is normally defined as consisting of a finite $m \times n$ array of cells, each of which can be in one of two possible states: alive represented by the number one, or dead represented by the number zero; on the computer screen, living cells are coloured black and dead cells are coloured white.

The state of the cells as time progresses is determined by the state of their eight nearest neighbouring cells. There are essentially four rules that determine the fate of the cells at the next tick of the clock:

- *Birth:* A cell that is dead at time t becomes alive at time $t + 1$ if exactly three of its neighbours are alive at time t.
- *Death by overcrowding:* A cell that is alive at time t will die at time $t + 1$ if four or more of its neighbours are alive at time t.
- *Death by exposure:* A cell that is alive at time t will die at time $t + 1$ if it has one or no live neighbours at time t.
- *Survival:* A cell that is alive at time t will remain alive at time $t + 1$ only if it has either two or three live neighbours at time t.

Whilst the environment, represented as E, is defined as the number of living neighbours that surround a particular live cell, a fertility coefficient, represented as F, is defined as the number of living neighbours that surround a particular dead cell. Note that both the environment and fertility vary from cell to cell and indeed from time to time as the automaton evolves. In this case, the life of a currently living cell is preserved whenever $2 \leq E \leq 3$ and a currently dead cell will be reborn whenever $3 \leq F \leq 3$.

Clearly, a number of alternative rules other than (2, 3, 3, 3) can be set. The general form for such rules is (E_{min}, E_{max}, F_{min} and F_{max}) where $E_{min} \leq E \leq E_{max}$ and $F_{min} \leq F \leq F_{max}$. The CAMUS implementation of the Game of Life algorithm enables the user to design rules beyond Conway's original rule (Figure 6.4).

Figure 6.4 Game of Life in action.

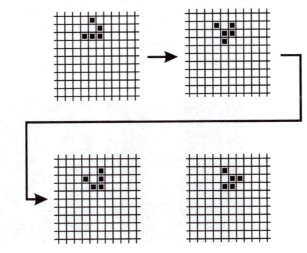

Figure 6.5 Examples of well-known initial configurations for the Game of Life that lead to interesting emerging behaviour.

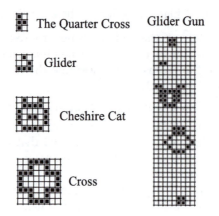

The original Game of Life setting is also characterised by a number of interesting initial cell configurations that have given rise to intriguing emergent behaviour (Figure 6.5). Coincidence or not, these seem to produce better musical results too.

6.1.2 Demon Cyclic Space

The cells of the Demon Cyclic Space automaton assume more than two states. Each of the n possible states is represented by a different colour and they are numbered from 0 to $n - 1$. The evolution rules operate as follows: a cell that happens to be in a certain state k at one tick of the clock dominates any adjacent cells that are in state $k - 1$, meaning that these adjacent cells change from $k - 1$ to k. This rule resembles a natural chain in which a cell in state two can dominate a cell in state one even if the latter is dominating a cell in state zero. But here the chain has no end because the automaton is cyclic; that is, a cell in state zero dominates its neighbouring cells that are in state $n - 1$. This CA generates miniature worlds of incredible complexity. Initialised as a random distribution of coloured cells, it always ends up with stable, patchwork-type patterns, reminiscent of crystalline growths (Figure 6.6).

Figure 6.6 Demon Cyclic Space in action.

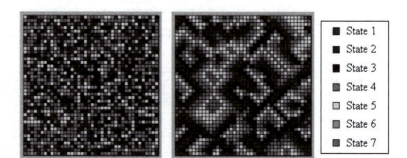

Figure 6.7 A toroidal space is normally employed to implement a cellular automaton.

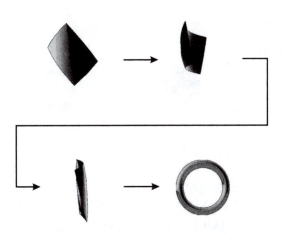

Both the Demon Cyclic Space and the Game of Life cellular automata function in a *toroidal space*: the right edge of the two-dimensional grid of cell wraps around to join the left edge and the top edge wraps around to join the bottom edge (Figure 6.7).

6.1.3 A cellular automata musical engine

CAMUS uses a Cartesian model in order to represent a triple of notes. In this context, a triple is an ordered set of three notes that may or may not sound simultaneously. These three notes are defined in terms of the distances between them, or intervals, in music jargon. The horizontal co-ordinate of the model represents the first interval of the triple and the vertical co-ordinate represents its second interval (Figure 6.8).

Figure 6.8 CAMUS uses a Cartesian model to represent a triple.

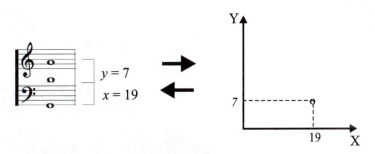

To begin the music generation process, both CA are set up with an initial random configuration and are set to run. When the Game of Life automaton produces a live cell, its co-ordinates are taken to estimate the triple from a given lowest reference note (Figure 6.9). This note is picked from a set defined by the user beforehand and a number of probability functions can be used for selection. Although the cell updates occur in parallel at each

Figure 6.9 Each screen of the Game of Life automaton produces a number of triples.

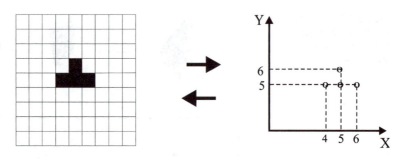

time step, CAMUS plays the live cells column by column, from top to bottom, before letting the CA update.

CAMUS uses both automata in parallel to produce music. The Game of Life automaton produces triples and the Demon Cyclic Space automaton determines the orchestration of the composition. The user allocates the colours associated to the states of the CA to MIDI instruments, so that the Demon CA designates the instrument that will perform the notes generated by a cell in the same (x, y) position of the respective Game of Life cell.

Figure 6.10 shows an example involving both CA. In this case, the cell in the Game of Life at position (5, 5) is alive and will thus generate a triple of notes. The corresponding cell in the Demon Cyclic Space is in state 4, which means that the sonic event will be played by the MIDI instrument in channel four. The co-ordinates (5, 5) describe the intervals of the triple: a fundamental pitch is given, then the next note will be at five

Figure 6.10 CAMUS uses two cellular automata in parallel to produce music.

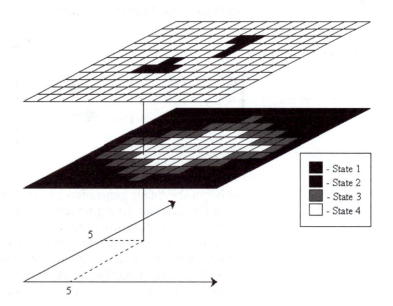

semitones above the fundamental and the last note ten semitones above the fundamental.

The notes within a cell can be of different lengths and can be triggered at different times. Once the triple of notes for each cell has been determined, the states of the neighbouring cells in the Game of Life are used to calculate a timing template, according to a set of temporal codes (Figure 6.11). These codes establish an abstract shape for the triple; the absolute triggering time and length values are calculated separately by another subroutine of the program (Figure 6.12).

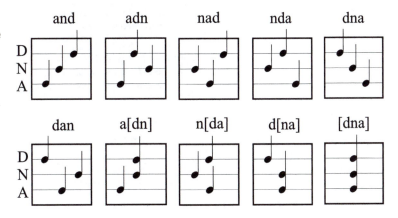

Figure 6.11 CAMUS uses timing templates to determine the rhythm of the triples.

McAlpine and Hoggar introduced a number of variations to the original program, notably the use of three-dimensional CA (Figure 6.13) and the use of Markov chains to generate the rhythms (Chapter 3). Both CAMUS and CAMUS 3D for Windows, plus examples and tutorials are available on the accompanying CD-ROM. McAlpine also performed a number of systematic experiments, which are documented in his doctoral thesis (McAlpine, 1999), whose relevant excerpts can also be found on the accompanying CD-ROM.

Also on the CD-ROM, in the folder **various\compositions**, there are excerpts in MP3 format from two pieces by this author composed using material generated by CAMUS and CAMUS 3D: *Entre o Absurdo e o Mistério* (for chamber orchestra) and *Grain Streams* (for piano and electro-acoustics).

6.2 Genetic algorithms

Genetic algorithms (also referred to as evolutionary algorithms) comprise computing methods inspired by biological processes,

Figure 6.12 Overview of the CAMUS processing.

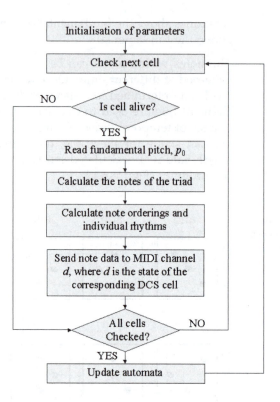

Initialisation of parameters

Check next cell

Is cell alive?

NO

YES

Read fundamental pitch, p_0

Calculate the notes of the triad

Calculate note orderings and individual rhythms

Send note data to MIDI channel d, where d is the state of the corresponding DCS cell

All cells Checked?

NO

YES

Update automata

Figure 6.13 The three-dimensional CA generates ordered sets of four notes per cell.

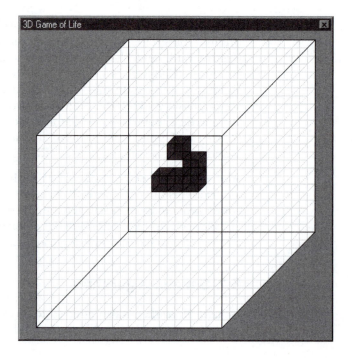

most notably those processes that are believed to be the driving forces of the origins and evolution of species, as proposed by Charles Darwin in the 1850s (Darwin, 1859) and more recently Richard Dawkins (1986). As in the case of cellular automata and artificial neural networks, genetic algorithms also embody a number of metaphors and jargon drawn from biology.

Genetic algorithms are normally employed to find optimal solutions to engineering and design problems where many alternative solutions may exist, but whose evaluation cannot be ascertained until they have been implemented and tested. In such cases, the optimal solution can be sought by manually generating, implementing and testing the alternatives or by making gradual improvements to alleged non-optimal solutions. As the complexity of the problem in question grows, however, these procedures become increasingly unsatisfactory and wasteful. Hence the need for an automated method that explores the capability of the computer to perform voluminous combinatorial tasks. Notwithstanding, genetic algorithms go beyond standard combinatorial processing as they embody powerful mechanisms for targeting only potentially fruitful combinations. These mechanisms resemble those of biological evolution such as natural selection based on fitness, crossover of genes, mutation and so forth; hence the label 'genetic algorithms'. (The term 'genetic programming' is sometimes used when the target solution is a program rather than data. In fact, genetic programming is sometimes thought of as a different field of research altogether.)

The sequence of actions illustrated in Figure 6.14 portrays a typical genetic algorithm. At the beginning, a population of abstract entities is randomly created. Depending on the application, these entities can represent practically anything, from the fundamental components of an organism, to the commands for a robot or the notes of a musical sequence. Next, an evaluation procedure is applied to the population in order to test whether it meets the objectives to solve the task or problem in question. As this initial population is bound to fail the evaluation at this stage, the system embarks on the creation of a new generation of entities. Firstly, a number of entities are set apart from the population according to some prescribed criteria. These criteria are often referred to as *fitness for reproduction* because this sub-set will undergo a mating process in order to produce offspring. The fitness criteria obviously vary from application to application but in general they indicate which entities from the current generation work best. The chosen entities are then combined (usually in pairs) and give birth to the offspring. During this reproduction process, the formation of the offspring involves a mutation process. Next, the offspring are inserted in the population. The fate of the remaining entities of the population not selected for reproduction may vary, but they usually die out without causing any effect. At this point we say that a new generation of the population has evolved. The evaluation procedure is now applied to the new generation. If the population still does not meet the objectives, then the system embarks once more on the creation of a new generation. This cycle is repeated until the population passes the evaluation test.

In practice, genetic algorithms normally operate on a set of binary codes that represent the entities of the population. These operations involve three basic classes of processes: *recombination, mutation* and *selection*. Whilst the recombination process causes the exchange of information between a pair of codes, the mutation process alters the value of single bits in a code. Recombination produces offspring codes by combining the information contained in the codes of their parents. Depending on the form of the representation of the codes, two types of recombination can be applied: real-valued recombination or binary-valued crossover.

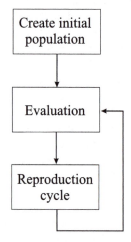

Figure 6.14 A typical genetic algorithm arrangement.

In order to illustrate a typical genetic algorithm in action, consider a population P of n entities represented as 8-bit codes as follows: $P = \{11010110, 10010111, 01001001, ...\}$. Then, suppose that at a certain point in the evolutionary process, the following pair of codes is selected to reproduce: $p_7 = 11000101$ and $p_{11} = 01111001$. Reproduction in this example is defined as a process whereby the couple exchange the last three digits of their codes followed by a mutation process. In this case, the pair p_7 and p_{11} will exchange the last three digits of their code as follows:

p_7: 11000[101] → 11000[001]
p_{11}: 01111[001] → 01111[101]

Next, the mutation process takes place; mutation usually occurs according to a probabilistic scheme (Chapter 2). In this example, a designated probability determines the likelihood of shifting the state of a bit from zero to one, or vice-versa, as the code string is scanned. Mutation is important for introducing diversity in the population, but one should always bear in mind that higher mutation probabilities reduce the effectiveness of the selective process because they tend to produce offspring with little resemblance to their parents. In this example, the third bit of p_7 and the fourth bit of p_{11} were mutated:

p_7: 11[0]00001   →   11[1]00001
p_{11}: 011[1]1101   →   011[0]1101

The new offspring of p_7 and p_{11} are 11100001 and 01101101, respectively.

6.2.1 Codification methods

In order to make effective use of genetic algorithms one has to devise suitable methods both to codify the population and to associate the behaviour of the evolutionary process with the application domain, which in our case is music. The rule of thumb is that one should try to employ the smallest possible coding alphabet to represent the population.

A typical codification method is the *binary string* coding whereby each individual is represented by a string of some specified length; the 8-bit coding illustrated above is a typical example of the binary string coding. This coding method is interesting because each digit of the code, or groups of digits, can be associated with a different attribute of the individual (Figure 6.15).

A number of variations of the binary string coding may be devised. For example, one could devise codes using large binary strings divided into words of a fixed length; each word would

Figure 6.15 Each digit of the binary string can be associated to a different attribute of the individual being coded; e.g., the attributes of a musical note.

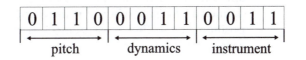

Figure 6.16 Each digit of the decimal string is computed at the binary representation level.

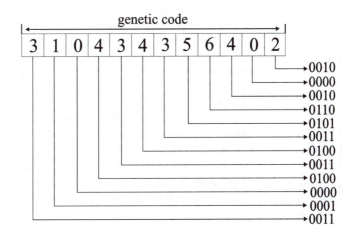

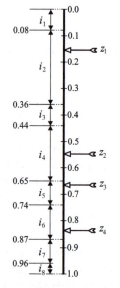

Figure 6.17 In stochastic sampling, a population is projected onto a line formed by segments of various lengths. Each individual in the population is assigned a segment whose length represents its fitness value.

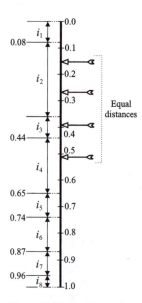

Figure 6.18 A variation of the stochastic sampling selection uses a metric template.

correspond to a decimal value. In this case, the code is a decimal string but the actual computing of the genetic algorithm processes occurs at their corresponding binary representation (Figure 6.16).

6.2.2 Selection mechanisms

The mechanism for selecting the population subset for reproduction varies according to the application of the genetic algorithm. This mechanism generally involves the combination of a fitness assignment method and a probability scheme (Chapter 3). One of the simplest selection mechanisms is the *stochastic sampling selection*. In order to visualise how this selection mechanism works, imagine that the whole population is projected onto a line made of continuous segments of variable lengths. Each individual is tied to a different segment whose length represents its fitness value (Figure 6.17). Then, random numbers are generated whose maximum value is equal to the length of the whole line and those individuals whose segments cover the values of the random numbers are selected.

A variation of the stochastic sampling selection would be to superimpose some sort of metric template over the line. The number of measurements on the template corresponds to the amount of individuals that will be selected and the interval is defined as $d = 1/n$, where n is the number of measurements. The position of the first measurement is randomly decided in the range of $1/n$.

Another widely-used selection mechanism is the *local neighbourhood selection*. In this case, individuals are constrained to interact only within the scope of a limited neighbourhood. The neighbourhood therefore defines groups of potential parents. In order to render this process more effective, firstly the algorithm selects a group of fit candidates (using stochastic sampling, for example) and then a local neighbourhood is defined for every candidate. The mating partner is selected from within this neighbourhood according to its fitness criteria. The neighbourhood schemes that are commonly used are ring, two-dimensional and three-dimensional neighbourhoods (Figure 6.19), but more complex schemes may also be devised. The distance between neighbours does not necessarily need to be equal to one, and not all immediate neighbours need to be considered (Figure 6.20).

Figure 6.19 Ring neighbourhood with distance = 3.

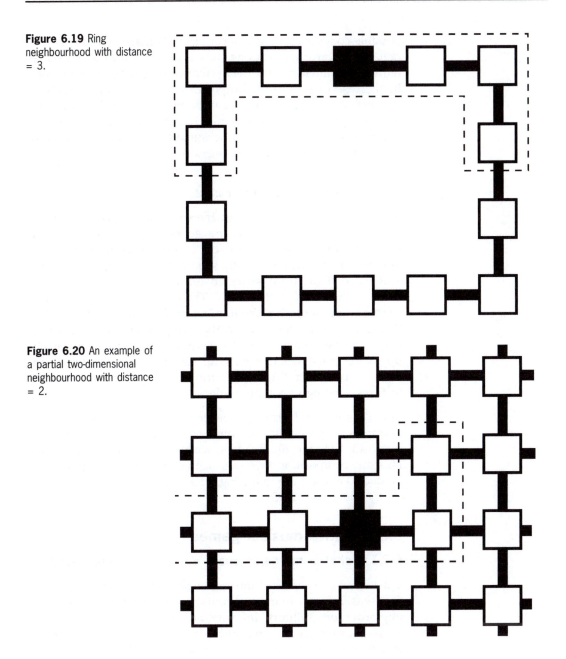

Figure 6.20 An example of a partial two-dimensional neighbourhood with distance = 2.

Many more selection methods exist but it is impracticable to survey all of them here. Successful applications of genetic algorithms often use selection mechanisms tailored specially to the task at hand.

Typically, genetic algorithms are used for optimisation purposes. Thus the selection operation is responsible for choosing the best possible codes according to a certain predetermined *fitness crite-rion*. There are applications, however, in which the user is required to interact with the system during the search process. In the case of a musical composition program, for example, a number of musical passages may be presented to the user, who then assigns fitness values of rankings to them based either on aesthetic judgements or on the subjective sense of how close these passages are to specific musical targets.

6.2.3 Growing musical organisms

By way of example we can cite the work of composer Gary Lee Nelson (Nelson, 1995) on using genetic algorithms to evolve rhythmic patterns. In this case, the binary-string method is used to represent a series of equally spaced pulses whereby a note is articulated if the bit is switched on (i.e. bit = 1) and rests are made if the bit is switched off. The fitness test is based on a simple summing test; if the number of bits that are on is higher than a certain threshold, then the string meets the fitness test. High threshold values lead to rhythms with very high density up to the point where nearly all the pulses are switched on. Conversely, lower threshold settings tend to produce thinner textures, leading to complete silence.

Vox Populi and Harmony Seeker on the CD-ROM are two examples of programs for composing with genetic algorithms (refer to Chapter 8). Readers wishing to learn more about musical composition systems using genetic algorithms are invited to consult the following references: Biles (1994), Degazio (1997) and Jacob (1995).

6.3 Adaptive musical games

6.3.1 A shot in the dark: where does music come from?

Music is one of the most intriguing phenomena of human kind. Our sensitivity to timing and imitation, our tendency for impos-ing order on auditory information, our ability to categorise sound, our tendency to recognise and imitate sound patterning, and so on, are all unique to humans (Storr, 1993). Babies, for example, have an extraordinary disposition to extract little chunks of sounds from the undulating patterns of speech and to imitate these sounds (Jusczyk, 1997). In most cases, in order to extract these little chunks of sounds, the left ear tends to seek timing cues in the signal; for instance, the begin and end of longer periodical signals (e.g., vowels) demarcated by short non-periodic bursts of energy (e.g., consonants). Conversely, the right ear tends to listen for sound coloration cues such as tone of voice and the melodic contour of utterances (Best, 1988). In the early months, the only response babies can make to heard utterances is to babble and they seem to respond better to clearly articulated syllables and exaggerated melodic contours. This is because they are still learning how to play with the raw building materials of music and language. Virtually all babies babble; even babies with auditory and/or oral difficulties seem to babble with their hands. During this learning process they develop their neural and muscular apparatus, firstly by learning how to *recognise* utterances and then by trying to *imitate* them. Preferences for listening to and imitating sounds in the mother tongue begin to emerge due to positive feedback responses that babies receive from their interlocutors. Sooner or later children start to form lexicons, to make sentences, to shape grammars, and to engage in increasingly sophisticated social interactions.

There is a school of thought that suggests that our musical predisposition provides the basis for the development of the sophisticated discrimination and categorisation machinery that are vital for language. However, our musicality, so to speak, certainly develops together with language as we grow. But how has this musical predisposition evolved in humans? And why? We are still groping in the dark, but there are some interesting conjectures, a couple of which are discussed below.

The notion that our linguistic capacity is closely related to our ability to both make and appreciate music was prominent in Enlightenment thinking in the eighteenth century. Reflections on how primordial utterances, cries and vocalisations would have evolved into language naturally brought musical considerations within the scope of the writings of philosophers such as Condillac and Rousseau (Thomas, 1995).

Condillac in his *Essay on the Origins of Human Knowledge* depicted the earliest spoken language as being composed of action-orientated vocal inflections such as warnings, cries for help, shouts of joy, and so forth. But most interestingly, Condillac proposed that these inflexions were accompanied by variations in pitch and timbre. In short, he suggested that early hominids did not prioritise the invention of different utterances, but tended to produce the same form of utterance at different tones in order to express different things; presumably by varying pitch, loudness and duration. Condillac thus suggested that primordial languages did not have many consonants but mostly vowel-like intonations. If we were to build on Condillac's conjecture, we could imagine that the prosody of earlier languages must have sounded like kinds of primitive songs.

Rousseau also purported a similar idea to Condillac. For Rousseau, however, song and speech have common ground: *passion*. In the beginning, vocal utterances primarily expressed feelings, whilst gestures were normally preferred to express rational thoughts. Rousseau agrees with Condillac that primeval spoken languages must have sounded like melodies of vowel-like utterances, but Rousseau has an interesting story for the emergence of consonants: as hominids' dealings with one another grew in complexity, spoken language needed to become less passionate and more precise. In his *Essay on the Origins of Language*, Rousseau argued that language was motivated by the increasing necessity for social bonding. Within this bonding process, the amount of tone variations decreased, giving rise to the appearance of consonants. New articulations needed to be formed, and consequently grammatical rules for making sequences of utterances soon emerged. For Rousseau, modern languages, including his own mother tongue French, no longer spoke to the heart alone, but also to reason. As language followed the path of logical argumentation, those melodic aspects of the primordial utterances evolved into music instead. Music thus, according to Rousseau, developed from the sounds of passionate speech.

Although dating back to the eighteenth century, these two philosophers' conjectures do continue to make sense to a certain extent (Brown, 2000; Richman, 2000). For instance, some non-Western languages, such as Cantonese, have mechanisms in which multiple intonations of the same word convey different meanings. This is perhaps an indication that Sino-Tibetan languages took a different evolutionary path from Indo-European ones. Another curious fact is that apparently the structure of the very early hominid's vocal tract (not necessarily our direct ancestors, but perhaps their cousins) could not produce plosive consonants such as /k/ (as for the word 'kit' in English) and /g/ (as for the word 'go') but it could produce vowels.

In the second half of the twentieth century, Wittgenstein (1963) in his *Philosophical Investigations* proposed a few notions that made a great impact on some veins of contemporary linguistic research and the echoes of this impact reached the ears of musicologists. Perhaps the most interesting of these notions is the notion of *language games*: simple linguistic plots specifically designed to illustrate particular points he wanted to make. Wittgenstein proposes a simple language game as follows: imagine a language that is meant to serve for communication between a builder and his assistant. The builder is constructing a house with building-stones: there are blocks, pillars, slabs and beams. The assistant has to pass him the stones, and in the order in which the builder needs them. For this purpose they use a language consisting of the words: 'block', 'pillar', 'slab' and 'beam'. The builder calls them out and the assistant brings the respective material he has learned to bring, in compliance to such-and-such a call. When the builder says 'block', for example, he causes an action to take place; the success of this action is conditioned to the fact that the assistant passes a block to the builder.

As far as the builder is concerned, it is sufficient that the assistant has learned only these four words without learning the whole English language. The point that Wittgenstein wanted to illustrate with this simple game is that, in principle, the assistant learned the language only by associating the sound patterns of the words uttered by the builder with actions, rather than with labelled pictures of different stones that may or may not appear catalogued in his mind. The indication that the assistant has learned the meaning of the words is given when he performs the right actions. To put it simply, Wittgenstein's point is that words (and sentences, for that matter) have meaning only when they have a role in specific contexts within the web of human activities.

Although Wittgenstein was not primarily concerned with the origins of music himself, his notions of *language games* and of the *action-based origins of meaning* are certainly inspiring. Coincidence or not, some of the evolutionary modelling techniques that have emerged from artificial life (alife) research are remarkably suitable for implementing Wittgenstein-like games within a musical context, as we shall see in the following sections.

6.3.2 Evolutionary music modelling

A plausible approach to alife-like musical models and simulations is to consider music as an adaptive system of sounds used by a number of individuals in a certain community, or *distributed agents* in computer science jargon, engaged in a collective music making experience. Some of these agents may only listen to the sounds while others may be fully engaged in the generative process. There should be no global supervision taking place in such collective activity and the agents must not have direct access to the musical knowledge of the other agents, apart from *hearing* what they actually produce during the interactions. As a rough analogy, this model could be metaphorically compared to a jam session where people, who have not necessarily met before, can join in and play. A contrasting scenario would be an orchestral concert where musicians follow a common score under the direction of a conductor; in this case we would say that there is almost no room for musical evolution while the orchestra is performing.

It is opportune to clarify at this stage the sense in which the notion of evolution should be taken in the context of evolutionary music modelling. Evolution is generally associated with the idea of the transition from an inferior species to a superior one and this alleged superiority can often be measured by means of fairly explicit and objective criteria: we believe, however, that this notion should be treated with caution. Indeed, this approach made a serious impact on cultural anthropology, which until recently maintained that human kind has become increasingly sophisticated during the linear course of history. The legacy of this argument is the popular general belief that the Stone Age, for example, represents far less sophistication than the Iron Age. This approach to evolution suffers from taking for granted that one can measure the current status of a cultural system entirely based on materialistic criteria. Alas, one should not take for granted that European classical music, for example, is more sophisticated than African drumming solely based on the technological sophistication of the instruments used. The relatively rudimentary technological development of most non-European societies gave rise to intricate social systems and complex religious rituals which led to rather sophisticated musical manifestations (Reck, 1997). A remote community living in an environment where prey and crops abound would not be under pressure to develop sophisticated hunting weaponry or irrigation technology, for instance. These people would probably give priority to the creation of a belief system in which a religious ritual would have to be performed occasionally in order to maintain the richness of their habitat.

With reference to prominently cultural phenomena, such as music, the notion of evolution surely cannot not have exactly the same connotations as it does in natural history: biological and cultural evolution are therefore quite different domains. Cultural evolution should be taken here as the transition from one state of affairs to another, not necessarily associated with the notion of improvement. Cultural transition is normally accompanied by an increase in the systems' complexity, but note that 'complex' is not a synonym for 'better'.

In the opening pages of the first issue of *Evolution of Communication Journal*, Luc Steels proposes four fundamental mechanisms for studying the origins and evolution of language:

(a) transformation and selection (originally termed 'evolution' by Steels), (b) co-evolution, (c) self-organisation and (d) level formation (Steels, 1997). As these mechanisms are strikingly suitable for the study of the origins and evolution of music as well, let us apply them from a musicological point of view as follows.

Transformation and selection

When a transformation process creates variants of some type of entity, there is normally a mechanism that favours the best transformations according to certain criteria. For example, the criteria in biology might be based upon some sort of endowment for the survival of an organism, whilst in linguistics the criteria might involve a compromise between the effortless use of the speech apparatus and the degree of understanding of an utterance. What is important here is that these transformations must *preserve the information* of the entity, otherwise the entity is destroyed rather than transformed.

In a way, the selectionist process of genetic algorithms, introduced earlier, can be regarded as an example of the transformation and selection process: selection here is based on the fitness for survival of a living organism represented in a virtual ecosystem. In music, however, there is no such a thing as a living organism battling for survival; the entities here are simply sounds produced by a community. Transformation and selection thus operate on these sounds and the selection process is more likely to involve criteria drawn from the psychological and physiological constraints of the individuals in the community that use these sounds, rather than basic biological needs such as mating or eating. (There are, however, instances where song is used to attract mates; birdsongs.)

Co-evolution

Co-evolution involves the interaction of various contiguous transformation and selection processes. Co-evolutionary criteria are not fixed, but rather they are affected by an environment that is also in a state of flux. In an artificial model, this can be implemented by replacing the fixed environment of evolution with another transformation and selection process so that it can also change. Whilst transformation and selection tend to drive a system towards the improvement of particular aspects, co-evolution tends to push the whole system towards greater complexity in a *co-ordinated manner*.

As a rough example, consider the case of the keyboard class of instruments: the evolution of the piano is popularly associated with the settlement of the equal-temperament tuning system (a tuning system in which all notes of the scale are equally separated by exactly half a tone) and with the increasing use of expressive dynamics in compositions; the piano can produce a much wider band of variations in dynamics than the harpsichord, from extremely loud to extremely quiet notes. The equal-temperament system alleviated the problem of tuning different instruments in an ensemble, but it also increased the complexity of the compositional praxis, as it augmented the possibilities for making use of more complex harmonic structures (e.g., the easy modulation to different tonal keys). We can tentatively conclude that new musical styles co-evolved with the evolution of musical instruments, and vice-versa. We say 'tentatively' because it is not entirely clear what aspects of musical style led to the development of the piano. Selective processes need to operate on the sides of both technology and style, but much more research is needed to identify what these processes actually are.

Self-organisation

The notion of self-organisation is closely related to the notion of *coherence*. The emergence of coherence in distributed systems with many interacting agents is a vast research topic. To put it in simple terms, three ingredients are needed for self-organisation to take place in a system: (a) a set of possible variations, (b) random fluctuations and (c) a feedback mechanism. Random fluctuations in the system will eventually strengthen some of the variations because of the feedback mechanism: the more a variation is strengthened, the more predominant it becomes. As an example, imagine the following scenario: a group of agents engages in a drumming session but none of the agents is an experienced musician; they have no musical training. Each agent can bring in a different percussion instrument but the agent must never have played the instrument before. They all agree that they should start playing sounds simultaneously in any way they wish. To begin with, they will certainly produce a highly disorganised mass of rhythms. The set of possible variations in this system is the set of all noises and rhythms that can be produced by the instruments. Next, imagine a situation in which at some point an agent A makes a sound pattern that catches the attention of a fellow agent B. As a consequence, the fellow agent attempts to imitate it. The imitation may not be exact (for example, because agent B is playing an instrument that is different from the one that agent A is playing), but agent A recognises it as an imitation of the pattern and instinctively reproduces the original pattern. In this case we would say that agent A gave a positive feedback to agent B. The other agents will probably be keen to imitate this pattern as well and variations will certainly start to emerge. After a while, the pattern that was originated by agent A, plus its variations, will become conventions. The next time these agents engage in a jam session they will certainly remember these and other patterns, and will probably play them during the session. The more the patterns are played, the more conventional they become. When not engaged in jam sessions, some agents may even consider adapting their instrument to better produce those patterns, whilst others will spend a great deal of effort practising ways to produce them.

Most cultural phenomena seem to follow identical self-organising principles in some way, but at different time-scales. For an example of self-organising principles operating in a time-scale longer than that of a few jam sessions, refer to an interesting account of the origins of the flamenco song style in a paper by Washabaugh that appeared in the *Journal of Musicological Research* (1995).

Level formation

Level formation involves the formation of *higher-level compositional conventions*. Suppose that at some point in the musical scenario described above, agents start remembering rhythmic patterns in terms of repeated short sequences grouped together as units; for example, shorter sounds may be grouped by similarity in duration and proximity in time, and repeated patterns may be grouped as larger units and so forth. This figurative conceptualisation of rhythm should then yield more abstract conceptualisations such as metric rules and a sense of hierarchical functionality.

6.4 Evolving rhythmic forms

This section presents an evolutionary music model to study the emergence of rhythmic forms in a virtual community of agents engaged in a collective music making activity. The usefulness of such a type of model for composition will be discussed later. The model is inspired by Wittgenstein's notion of language games, by Steels' evolutionary mechanisms and by the recognition-by-imitation hypothesis whereby the tight coupling of recognition and imitation plays a key role in the development of the musical ability of infants.

In the context of this experiment, it is assumed that rhythmic forms result from the emergent behaviour of *interacting distributed agents*. The motivation behind these interacting agents is to produce a shared repertoire of short rhythmic patterns. The model should comply with two fundamental requirements in order to produce realistic results. Firstly, there must be no global mechanism or rules controlling the interactions or coaching the agents towards the formation of the rhythmic repertoire. The agents must be autonomous in the sense that they interact with one another solely based upon their own internal rules. In computer science jargon we say that the control is distributed, hence the term 'interacting distributed agents'. Secondly, an individual agent can only gather information from what is externalised by another agent. That is, agents cannot tap into the memory of other agents in order to see what is in there.

Before we go on to describe the anatomy of the agents, let us briefly examine what an external observer would see by watching these agents in action. Different from the example given above to illustrate self-organisation, the interactions here occur on a two-by-two basis: a pair of agents is randomly selected for interaction and one of them (agent A) plays a rhythmic pattern to the other (agent B). The task of agent B is to try to imitate agent A. Due to the fact that the agents can only understand what they hear in terms of the rhythmic patterns that they already know, agent B must play back a pattern taken from its repertoire. If its repertoire is empty, then it makes a response at random and plays it. If, on the other hand, its repertoire is not empty, then it carefully listens to agent A and selects the most similar pattern that it can find in its repertoire for the imitation. Next, agent A assesses whether or not the imitation is acceptable and gives feedback to agent B. If the imitation was acceptable then agent A reproduces the original pattern, but if it was unacceptable, then it just shuts up and silently terminates the interaction. After this interaction, both agents inspect their memories and perform a few updates based upon the experience they both had. The cycle recommences, and so on. Depending upon the size of the population, the nature of the agents and other modelling constraints, thousands of rounds might be necessary until a shared repertoire of rhythmic patterns starts to emerge.

Note that to begin with, the agents have no repertoire in their memory. The consequence of not having a repertoire for an agent are:

1   If the agent sets itself to play something, then it will have to create a new pattern from scratch.

2   If the agent hears a pattern, it will not be able to infer anything from it; that is, the agent cannot actually listen to what it hears. In cognitive science parlance, we could say that an agent with an empty repertoire is unable to perceive and enact. This is a consequence of the recognition-by-imitation hypothesis mentioned earlier.

An important presupposition of our model is that whilst the action of playing a rhythmic pattern involves the activation of certain motor mechanisms in specific ways, the recognition of a rhythmic pattern requires knowledge of the activation of these motor mechanisms: that is, which mechanisms would have to be activated and how, in order to reproduce the pattern in question. It does not matter, however, if one agent pulls different strings to another in order to produce similar patterns.

6.4.1 The anatomy of the agents

Each agent is composed of: (1) a voice synthesiser, (2) a hearing apparatus, (3) a memory device and (4) a cognitive module. The *voice synthesiser* implements a source/filter architecture (refer to Figure 7.25 in Chapter 7) combined with a geometric articulator that allows for control using parameters such as tongue position, lips and jaw opening. Basically, the synthesiser requires 12 parameters to produce a sound:

f_0 = pitch
w_1 = amplitude of the voiced source signal
w_2 = amplitude of the white noise source signal
z = nasal cavity switch (or velic closure)
a = horizontal position of the tongue's body
b = vertical position of the tongue's body
c = rounding of the lips
f_m = vibrato width
d = vibrato depth
j = jitter coefficient
h = duration
t = articulatory clock

An introduction to voice synthesis and to the source/filter architecture is given in Chapter 7, but it is not necessary to study the role of the entire set of parameters listed above in order to understand how our model works. It suffices to understand that in order to synthesise a rhythmic pattern, an agent has to compute a sequence of articulatory commands that will be triggered according to the clock t:

(synthesise f_0 w_1 w_2 z a b c f_m d j h t^0)
(synthesise f_0 w_1 w_2 z a b c f_m d j h t^1)
(synthesise f_0 w_1 w_2 z a b c f_m d j h t^2)

...

(synthesise f_0 w_1 w_2 z a b c f_m d j h t^n)

Thus, to synthesise the rhythmic pattern in Figure 6.21, the agent would have to produce something like this:

(synthesise 493.9 0.5 0.12 0 0.5 0.75 0.25 5.6 0.016 0 1.0 0.0)
(synthesise 493.9 0.5 0.12 0 0.5 0.75 0.25 5.6 0.016 0 1.0 1.0)
(synthesise 000.0 0.0 0.00 0 0.5 0.75 0.25 5.6 0.016 0 1.0 2.0)
(synthesise 493.9 0.8 0.20 0 0.5 0.75 0.25 5.6 0.016 0 1.0 3.0)

Figure 6.21 An example of a simple rhythmic pattern.

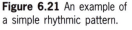

As for the *hearing apparatus*, it performs a *time-onset* analysis on an incoming signal. Time-onset analysis works by looking for peaks of high energy in the topology of the sound in the time-domain. The analyser records the peak amplitude values and the time they occur (Figure 6.22).

Figure 6.22 The time-onset analysis looks for peaks of high energy in the topology of the sound in the time-domain.

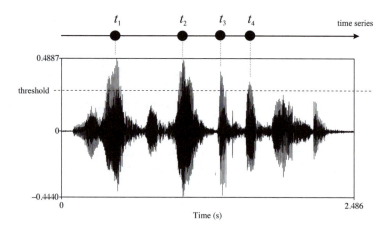

The *memory device* stores the evolved rhythmic repertoire and other data such as probabilities, thresholds and reinforcement parameters. The method for representing this knowledge is introduced below. A sample of the parameters that each agent has to manage is given as follows (their meaning will be clarified later):

$$
\begin{aligned}
n &= \text{deviation coefficient} \\
d &= \text{distinctiveness threshold} \\
Q &= \text{time resolution} \\
A &= \text{loudness resolution} \\
P_a &= \text{creative willingness} \\
P_b &= \text{forgetfulness disposition} \\
r &= \text{rounds threshold} \\
s &= \text{reinforcement threshold}
\end{aligned}
$$

Finally, the *cognitive module* provides the agent with the knowledge on how to behave during the collective musical activity. For example, the agent must know when to remain quiet, what to do when the player produces a pattern, how to assess an imitation, and so forth. By and large, it is the cognitive module that determines the dynamics of the interactions.

For the sake of simplicity, the cognitive module does not evolve in the present model. Metaphorically speaking we could say that the agents were born with this module hard-wired in their brains. Moreover, all agents are alike with respect to their cognitive module; that is all of them follow the same procedures. Also, all agents have identical synthesis and listening apparatus that do not evolve. In other words, all agents are clones of each other in this model. A more realistic model would certainly need, however, to incorporate agents with varied abilities and with a changeable cognitive module. The following section presents an informal description of the knowledge held by this module.

6.4.2 The cognitive module's knowledge

As mentioned earlier, the interactions between the agents occur on a two-by-two basis. The goal of the agents is to develop a shared repertoire of rhythmic patterns and to imitate each other as well as possible.

At each round, both agents in a pair chosen from the community play two different roles: the player and the imitator. Once two agents have been chosen for a round, one of them takes the role of the player by producing a pattern p_1, randomly chosen from its repertoire. If its repertoire is empty, then it produces a random pattern. The other agent then assumes the role of the imitator: it analyses the pattern p_1, searches for a similar pattern in its repertoire, p_2, and produces it. The player in turn analyses the pattern p_2 and compares it with all other patterns in its own repertoire. If its repertoire holds no other pattern p_n that is more similar to p_2 than p_1 is, then the player replays p_1 as a reassuring feedback for the imitator; in this case the imitation would be acceptable. Conversely, if the player finds another pattern p_n that is more similar to p_2 than p_1 is, then the imitation is unsatisfactory and in this case the player would terminate the interaction without emitting the reassuring feedback. It should be noted that in all these cases, the agents never play a pattern exactly as it is represented in their memory: they always add a small deviation to its realisation (the amount of deviation is determined by the coefficient n).

If the imitator hears the reassuring feedback, then it will reinforce the existence of p_2 in its repertoire and will change it slightly as an attempt to make the pattern even more similar to p_1. In practice, the reinforcement is implemented as a counter that registers how many times a pattern has successfully been used. Conversely, if the imitator does not receive the feedback then it will infer that something went wrong with its imitation. In this case, the agent has to choose between two further courses of action. If it finds out that p_2 is a weak pattern in its memory, because it has not received enough reinforcement in the past, then it will try to modify p_2 slightly, as an attempt to further approximate it to p_1. It is hoped that this approximation will give the pattern a better chance of success if it is used again in another round. But if p_2 is a strong pattern, then it is because it has been used successfully in previous imitations and a few other agents in the community also probably know it. In this case, the agent will leave p_2 untouched, will attempt to create a new pattern similar to p_1, and will include this new pattern in the repertoire. The creation of this new pattern is done on a trial-and-error basis: the agent 'babbles' for a few trials and selects the one that is most similar.

Before terminating the round, both agents perform final updates. These updates are: *merge*, *forget* and *create*. Firstly they scan their repertoire and merge those patterns that are considered to be too similar to each other. Also, at the end of each round, both agents have a certain P_b probability to undertake a spring cleaning to get rid of weak patterns: if a pattern has been used for more than r rounds and if this pattern has led to less than $s\%$ of satisfactory imitations, then it is deleted from the memory. In other words, those patterns that are not sufficiently reinforced in the memory are simply forgotten. Finally, at the end of each round, both agents have a certain P_a probability of adding a new randomly created pattern to their repertoires.

In practice, all decisions involving pattern similarity, whether for assessing an imitation, for creating a similar pattern or for deciding whether or not to merge two patterns, are made based upon distance measurements. Regarding the merge update, there is the distinctiveness threshold value, *d,* whereby patterns are merged if the distance between them is smaller than this threshold.

In order to gain a better understanding of how the agents compare and handle the rhythmic patterns, let us briefly examine how these patterns are represented in the memory of the agents.

6.4.3 The memorisation of rhythmic patterns

The agents store rhythmic information in terms of *binary and ternary onset strings.* As far as the agents are concerned, these are the fundamental building blocks for making larger and more elaborate rhythmic patterns; they can be combined to form strings of four, five, seven, eight onsets, and so on (Figure 6.23).

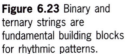

Figure 6.23 Binary and ternary strings are fundamental building blocks for rhythmic patterns.

These fundamental building blocks are represented in terms of the *loudness* (or amplitude) and *timing* of the individual onsets. Note that the representation of the loudness and timing values are not absolute, but proportional. Let us take as an example the binary onset string shown in Figure 6.24: the loudness of the first onset is 30% higher than the loudness of the second onset, in relation to a maximum default value. The distance between onsets *a* and *b* is 50% of the total length of the string. The agents represent binary onset strings as points in a Cartesian space whose co-ordinates are given by these percentage values. For didactic reasons, from now on we will focus only on binary onset strings; the ternary ones are represented in a similar fashion.

The distance between two binary strings is computed as the Euclidean distance between two points in a Cartesian space:

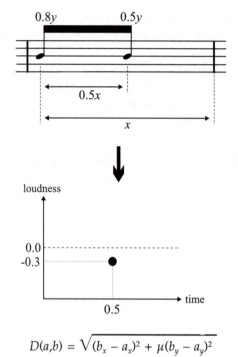

Figure 6.24 The representation of a binary onset string as a point in a Cartesian space. The x-axis is given by the proportional onset distance and the y-axis is given by the proportional difference in loudness. These values are represented in the graph as decimals.

$$D(a,b) = \sqrt{(b_x - a_x)^2 + \mu(b_y - a_y)^2}$$

(The coefficient μ is used to differentiate the physical nature of the two dimensions (i.e., time and amplitude) but for the sake of simplicity, it will not be taken into account in the following examples.) Figure 6.25 illustrates three binary onset strings and the respective distances between them.

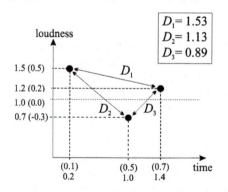

Figure 6.25 Measuring the distances between binary onset strings. Note that the original values of the co-ordinates (in parenthesis) have been scaled in order to calculate the distances: time values were multiplied by two and loudness values were offset by 1.0.

We are now in a position to introduce parameters Q and A listed previously: the time and loudness resolutions, respectively. These two parameters define the granularity of the Cartesian space. The higher the granularity, the higher the number of distinctive patterns that the agents can handle; this difference is illustrated in Figure 6.26.

Figure 6.26 The two distinct patterns in the space on the left side cannot be distinguished from the lower resolution of the space on the right.

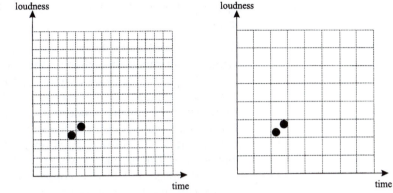

6.4.4 Examples

In order to illustrate the dynamics of our model, let us examine two interaction examples. The first example illustrates a case where the player has only one binary onset string in its repertoire whereas the imitator has three (Figure 6.27). The player starts by playing the only string it knows: p_1. The imitator analyses p_1 in order to infer where it would stand in its Cartesian space and compares the distances between p_1 and its three strings i_1, i_2 and i_3. The imitator finds out that the shortest distance is between p_1 and i_2, and so it produces i_2 as an imitation of p_1. The speaker analyses i_2 in order to infer where it would stand in its Cartesian space and compares the distances between i_2 and the strings of its repertoire. Since there is only one string in the repertoire of the player, i_2 is considered to be acceptable imitation; even though it might sound very different to an external observer. The player gives a reassuring feedback to the listener, who in turn reinforces the strength of i_2 in its repertoire and shifts its representation slightly towards p_1. Both

Figure 6.27 A case where the imitation is successful. Even though the imitation may sound rather different to an external observer, the agent is not able to fully judge the imitation due to the lack of variety in its repertoire.

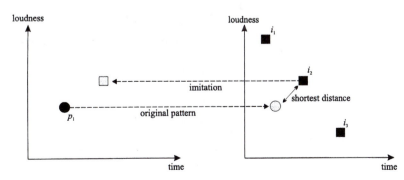

agents perform the halting updates and the interaction terminates.

The next example illustrates a case where the player has four
strings in its repertoire whereas the imitator has three (Figure
6.28). The player starts by playing p_3 and the imitator analyses
it in order to infer where it would stand in its Cartesian space.
It compares the distances between p_3 and its two strings i_1 and
i_2. The imitator finds out that the shortest distance is between p_3
and i_1, and so it produces i_1 as an imitation of p_3. The player
analyses i_1 in order to infer where this string would stand in its
Cartesian space and then it compares the distances between i_1
and each of the strings of its repertoire. Unfortunately the player
finds out that the shortest distance is between i_1 and p_1. The
imitation i_1 is therefore closer to p_1 than to p_3: in this case this
imitation is not acceptable. No reassuring feedback is given and
since i_1 has been successfully used in previous interactions, the
imitator leaves i_1 as it is, creates a new string from scratch and
adds this new string to its repertoire. This new creation should
be similar to p_3. Both agents perform the halting updates and the
interaction terminates.

Figure 6.28 A
case where the
imitation is not
satisfactory because
the player found out
that the imitation is
closer to another
string in its
repertoire than to
the one it had
produced in the first
place.

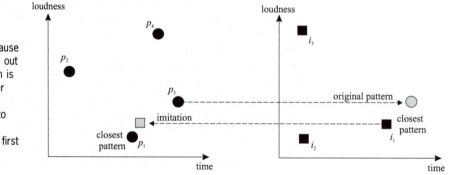

The graph in Figure 6.29 shows the repertoire of a community
of 20 agents, plotted on top of each other, after 4000 rounds of
interaction using the following settings:

n = 2%
d = 0.66
Q = 0.01
A = 0.01
P_a = 10%
P_b = 1%
r = 5
s = 80%

Figure 6.29 An evolved repertoire of binary onset strings.

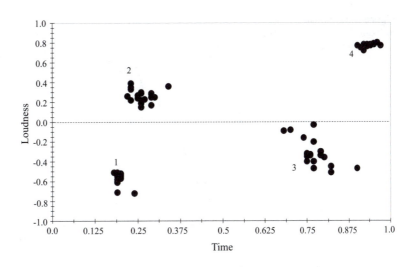

Note that a shared repertoire of binary onset strings has indeed emerged in the community. The clusters in the graph indicate that various agents share similar strings. They are not exactly equal because the agents have a certain tolerance for wandering within the clustering area. Examples of possible binary strings that are recognisable by the agents are shown in Figure 6.30. Note, however, that standard musical notation does not explicitly represent the tolerance within the clustering area. For example, cluster number two in Figure 6.29 embraces second onsets whose loudness (or dynamics, in music jargon) can be anything from *piano* to *mezzo forte* in relation to the loudness of the first onset (it is assumed that the value 0.0 on the *y*-axis of the graph corresponds to *pianissimo*, 0.2 to *piano*, 0.4 to *mezzo piano*, etc.). The duration of the onsets also suffers here as musical figures suggest an implicit quantisation that may distort the resolution of the Cartesian space. This tolerance is only implicit in the representation in the sense that performers tend to interpret symbols for dynamics as being relevant to specific contexts.

Figure 6.30 Examples of binary patterns that are recognisable by the community of agents.

The graph in Figure 6.31 plots the average size of the repertoire of the community taken after every 100 rounds. Note that after approximately 1800 rounds (marked by the vertical dashed line) the average size oscillates at around four strings per agent. This size limit is largely determined by the amplitude and time resolutions, A and Q, and also by the deviation coefficient n: the lower the resolution of the Cartesian space and the larger the value of the deviation coefficient, the smaller the size of the repertoire will be.

Figure 6.31 Plotting of the average size of the repertoire in the community.

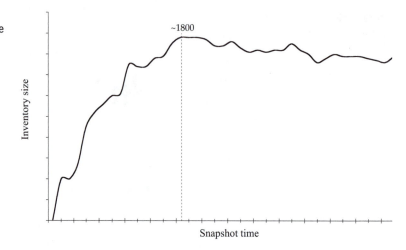

The graph in Figure 6.32 plots the histogram of the interactions in terms of the imitation success average taken every ten rounds. Here, the agents started with an empty repertoire but after all the agents had participated in at least one round, the average size of their repertoire soon settled to one, and remained so until the random creations took place according to probability coefficient P_a. Up to this point the imitations were invariably good. The most important thing that had happened by then was that the agents gradually shifted their strings in order to make them increasingly similar to each other; some clusters might have emerged at this point. Once random additions began to take place, more clusters emerged and indeed some agents occasionally had more strings in their repertoires than others. Imitations commonly failed and forced agents to add more strings to their

Figure 6.32 The histogram of the interactions in terms of imitation successes.

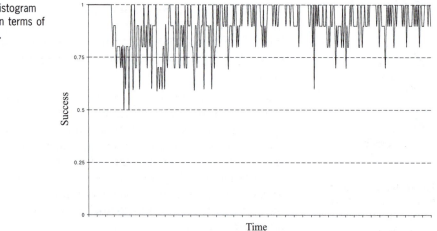

repertoires. New additions were quickly assimilated by the community, but after a certain number of additions the average size of the repertoire ceased to increase. Although at this stage the repertoire had stabilised, the strings were not static in the memory of the agents. Due to the fact that the agents always add some deviation to the actual realisation of the strings, the repertoire was in constant movement, but the clusters did not disperse. In some circumstances clusters might have merged or new ones might have appeared, but the average size of their repertoire tended to remain stable until the end of the simulation.

6.4.5 Concluding remarks

Roughly speaking, the settings of the model determine the *cognitive disposition* of the agents. Although the cognitive module fixes identical social protocols for the interactions between agents, some degree of variation can be introduced in this model by altering the cognitive disposition of the agents. In the case of the model presented here, this disposition does not vary during the interactions; it only can be manually changed beforehand. Nevertheless, the ability to change the cognitive disposition of the agents is crucial if such a model is coupled with other models, with the intention of initiating co-evolutionary and level-formation phenomena. For example, if for some reason a certain co-evolutionary process demands the creation of a larger repertoire of strings, then this process might compel the agents to increase the granularity of their time resolution. That is, the agents will be required to pay more attention in order to make more distinctions in the time domain. It could be argued that co-evolutionary mechanisms of this sort play a key role in the formation of groupings that may lead to the emergence of higher-level morphological tendencies. For example, the agents might, for some reason or constraint, prefer not to increase the repertoire of the basic binary and ternary onset strings, but rather to combine the ones they already know in order to form larger strings. Combinations of this sort should then lead to the emergence of a shared repertoire of rules for the formation of onset groupings and larger rhythmic patterns. Although no experimentation with co-evolutionary mechanisms has yet taken place with the present model, it is however possible to imagine the nature of such rules. For example, considering the evolved repertoire of fundamental building blocks in Figure 6.29, one of these potential rules could be grammatically defined as follows (refer to Chapter 2):

$$R \rightarrow b_1 P \mid b_2 Q$$
$$P \rightarrow b_1 P \mid b_1$$
$$Q \rightarrow b_2 Q \mid b_2$$

where:

P and Q are non-terminal nodes of the grammar;
$B = \{b_1, b_2, ...\}$ is the set of terminal nodes, each of which represents an onset string of the agent's repertoire;
R is the starting symbol of the grammar.

By way of an example, assume that the terminal node b_1 corresponds to the cluster number one in Figure 6.29. In this case, a rhythmic pattern generated by the above grammar can be given, as shown in Figure 6.33.

Figure 6.33 An example of a rhythmic pattern generated using evolved onset strings.

A similar adaptive game has been implemented by researchers at the Free University of Brussels, whereby the agents evolved repertoires of vowels rather than rhythmic building blocks. We conjecture that the rhythmic and the vowel models could be paired in a co-evolutionary system in order to forge the emergence of higher-level musical forms where time, loudness and formant frequencies contribute to the evolutionary phenomena. This is reminiscent of the co-operation between the left and right ears when processing the rhythm and the spectral contours of speech.

Would there not be a correspondence between the rhythms and musical scales of different cultures and the phonetic and phonological systems of their languages? Until more empirical evidence has been gathered, we can only speculate on the answers to this question. Nevertheless, the adaptive musical games approach to generating musical materials for pieces of music seems to be a sensible way forward in order to break new ground in algorithmic composition. This approach embodies mechanisms of cultural dynamics coupled with psychoacoustics and physical modelling: three spicy ingredients waiting for the right recipe. However, the model introduced in this section is rather simple, in the sense that the cultural dynamics presented in the examples may not require the complexity of the vocal and hearing apparatuses described earlier in order to produce these results: the examples have been kept simple for didactic reasons.

One of the main controversial aspects here is the fact that in reality, such iterations involve larger rhythmic patterns at each round and not merely binary or ternary strings. For instance, infants do not babble a single vowel or a short two-beat pattern; on the contrary, they babble fairly large streams of sounds. It would be foolish to consider that Condillac's or Rousseau's hominids would have evolved music or language by uttering only single syllables to each other. It could be argued that short chunks, or *memes* (Jan, 2000), are extracted from larger streams and processed in parallel, as if the agents were interacting by means of various concurrent rhythmic memes at each round. More experimentation is needed, however, to further the discussion. So far so good, this section on adaptive musical games should be regarded as a starting point for further investigation into this fascinating new area of research, which is likely to break new ground for musical composition practices.

6.5 Related software on the accompanying CD-ROM

There are three packages for evolutionary music on the CD-ROM: CAMUS (in the folder **camus**), Vox Populi (in the folder **populi**) and Harmony Seeker (in the folder **harmseek**). CAMUS is a program for generating music using cellular automata and it comes in two versions: CAMUS and CAMUS 3D. A hands-on tutorial and a reference manual are provided in the Help menu of both versions. Other documentation is also provided.

Vox Populi is a program for composing with genetic algorithms. Here, genetic algorithms are used to evolve sets of chords according to user-specified fitness criteria in terms of melodic, harmonic and voice range characteristics.

Finally, Harmony Seeker combines cellular automata and genetic algorithms to generate music. Basically, the program uses cellular automata to generate musical material that is scrutinised by a genetic algorithm whose function is to optimise the outcome of the system.

There is no software for composition available that embodies the adaptive games discussed in this chapter. These new ideas have only just begun to be systematically investigated and this book is probably the first to publish a discussion on such a topic. There is much research yet to be done before these concepts can be put into practice and evaluated. In the folder **various\evolution** there is a document in html format (to be read using a Web browser) containing an animation that complements the example given in Figure 6.29. A Nyquist implementation of the voice synthesiser used in the simulations is also available in this folder.

7 Case studies

This chapter is intended to give a glimpse into ways to put some of the techniques and concepts introduced in the previous chapters into practice. There are three short case studies. The first case study illustrates how one could start by using a combinatorial mechanism to produce musical material and use a rule base to give form to this material. The second case study illustrates a somewhat reverse process. Here a formal grammar is used first to generate melodic forms whose content is determined later by means of distribution functions.

Both case studies employ mechanisms that act upon sets of basic musical elements in order to generate musical content and rules to shape musical form. This is indeed standard practice in contemporary composition and it is ubiquitous in most algorithmic or automated composition systems. There is, however, one missing aspect here for which computers can provide excellent assistance, but it has been largely overlooked by composers: the definition of musical elements, or building blocks.

The third case study addresses this by proposing a method for deriving pitch systems from the inner structure of human vocal sounds. Here the computer is used both to dissect sound recordings in order to extract pitch information from them, and to synthesise this pitch information in order to monitor the process. In this case, synthesis can also play an active part in composition, as nowadays it is very common to compose with coexisting synthesised and acoustic sounds.

7.1 From content to form

The following paragraphs describe a composition method that uses a *combinatorial generative module* coupled with a set of *moulding rules*. The former generates raw compositional material and the latter moulds this material into musical passages according to a set of precepts. In this case study, the combinatorial module first generates miniature sequences of two, three and five chords, sorted into three different groups according to their length, and then the moulding rules turn these chords into musical forms and furnish them with musical attributes such as duration, timbre and dynamics. One could metaphorically regard the moulding rules as acting like an artisan giving form to raw pieces of clay (Figure 7.1).

Figure 7.1 The moulding rules give form to raw musical material.

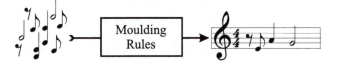

7.1.1 The combinatorial module

The mathematical basis of combinatorial processes was introduced in Chapter 2. Here, the functioning of the combinatorial module is illustrated by means of a practical example whereby it operates on a set containing nine different chords of four notes each, in context of the D major tonal key. The chords are as follows (Figure 7.2):

1 the tonic chord (represented as T)
2 the first inversion of the tonic chord (T_3)
3 the subdominant chord (S)
4 the first inversion of the subdominant chord (S_3)
5 the dominant chord (D)
6 the dominant chord with the seventh note (D7)
7 the first inversion of the dominant chord with the seventh ($D_3 7$)
8 the second inversion of the dominant chord with the seventh ($D_5 7$)
9 the third inversion of the dominant chord with the seventh ($D_7 7$).

Figure 7.2 The nine chords that are handled by the combinatorial module.

The task of the combinatorial module is to generate groups of two-, three- and five-chord strings. The module comprises three sub-modules: *centripetal*, *centrifugal* and *composite* sub-modules, each of which is responsible for each of the three groups of strings.

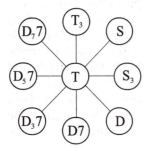

Figure 7.3 The centripetal generative chart.

The centripetal sub-module

The centripetal sub-module generates three-chord strings according to the chart shown in Figure 7.3. It generates the strings according to the following principles:

(a) The first and the last chord of the triple can be any of the peripheral elements of the chart but the middle chord must always be the central element; which in this example is the tonic chord (T).

(b) The central element must always be in the middle of the triple; it can never be the first or the last element of a triple. Examples of valid triples are: T_3-T-D7, D7-T-D7 and $D_5$7-T-S (Figure 7.4).

Figure 7.4 Examples of three-chord strings generated by the centripetal sub-module.

T_3 T   D7   D7   T   D7 $D_5$7   T   S

The centrifugal sub-module

The centrifugal sub-module generates two-chord strings according to the chart given in Figure 7.5. In this case, all chords are combined in pairs with the only restriction being that a chord cannot be paired with itself. For example, T-D7 and D7-T_3 are valid pairs, but D7-D7 is not allowed (Figure 7.6).

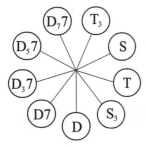

Figure 7.5 The centrifugal generative chart.

The composite sub-module

Finally, the composite sub-module combines triples and pairs randomly selected from the outcome of the centripetal and centrifugal modules, according to the following principles:

(a) Combinations can be in any order; that is, either a pair followed by a triple or a triple followed by a pair is allowed.

Figure 7.6 Examples of pairs generated by the centrifugal sub-module.

T   D7   D7   T

(b) The last chord of the first element must be different from the starting chord of the second element. For example, D-T-S + S_3-T is possible but T_3-$D_5$7 + $D_5$7-T-S_3 is not allowed because $D_5$7 appears both at the end of the first sequence and at the start of the next.

7.1.2 Sequencing the chords

So far the combinatorial module has generated three sets of short strings of two-, three- and five-chords, respectively. The next step is to build larger chord chains by sequencing these strings. As an example, let us compose a musical clip formed by three sections: a starting section containing six different centripetal triples, a middle section containing six different centrifugal pairs and a final section containing three different composite strings; these materials will be randomly selected from the respective outcomes. The crotchet will be adopted as a reference beat and the time signature will be set to four beats per bar. The metronome for the entire clip will be set to 60 beats per minute.

Figure 7.7 The overall form of the combinatorial musical clip.

60 M.M.

$\frac{4}{4}$	Section A	Section B	Section C
	6 centripetal triples	6 centrifugal pairs	3 composite streams
	semiquaver	crotchet	quaver

Next, let us assign a referential note duration figure for each section of the clip as follows: the semiquaver is allocated for the starting section, the crotchet for the middle section and the quaver for the final section. The overall form of the clip is portrayed in Figure 7.7. So far, the clip would resemble the score portrayed in Figure 7.8. The *modus operandi* for achieving this state of affairs is given by the following rules:

- *Metric Rule 1:* Place the chords sequentially within the bars according to the duration figure assigned to the respective sections.
- *Metric Rule 2:* If there is a change of section within a bar and the beat is not complete, then complete the current beat by enlarging the duration of the chord and start the next beat with the new corresponding duration figure (e.g., the transition from the first to the second beat in the second bar of Figure 7.8).

161

Figure 7.8 The musical clip in the making.

- *Metric Rule 3:* The duration of the last chord of the clip should last until the end of the last bar (e.g., the last chord of the seventh bar of Figure 7.8).

The next step is to orchestrate the material portrayed in Figure 7.8 and to create some form of articulation of this musical material. Let us establish that the musical clip will be orchestrated for six instruments, as follows (the notes of the chords are counted from the bottom up):

Instrument 1: Plays the top notes of the chords, transposed one octave higher.

Instrument 2: Plays the top notes of the chords (default option).

Instrument 3: Plays the third notes of the chords.

Instrument 4: Plays the second notes of the chords.

Instrument 5: Plays the bottom notes of the chords (default option).

Instrument 6: Plays the bottom notes of the chords, transposed one octave lower.

The orchestration is guided by the moulding rules discussed below.

7.1.3 The moulding rules

The following moulding rules take for granted that the vertical connections between the notes of the chords as seen in Figure 7.8 are unlocked. This is because these rules are intended to establish horizontal bonds and groupings between the notes in order to forge articulatory units, in addition to guiding the orchestration of the clip. The moulding rules are given as follows, in order of precedence:

- *Moulding Rule 1:* If two successive notes are equal in pitch, then they are tied in order to form one single note whose duration is equal to the sum of the duration of both. The resulting note constitutes an articulatory unit on its own; that is, it cannot be a member of articulations formed by other rules (Figure 7.9).

Figure 7.9 Moulding Rule 1 applied to the second notes of the chords in bar 3.

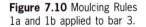

- *Moulding Rule 1a:* If Moulding Rule 1 is applied to the top notes of chords, then the result is played by Instrument 1 (Figure 7.10).

Figure 7.10 Moulcing Rules 1a and 1b applied to bar 3.

- *Moulding Rule 1b:* If Moulding Rule 1 is applied to the bottom notes of the chords, then the result is transposed down by one octave and played by Instrument 6 (Figure 7.10).
- *Moulding Rule 2:* All ascending sequences (see Rule 7) of notes are slurred in order to form an articulatory unit (Figure 7.11).

Figure 7.11 Moulding Rule 2 applied to the second notes of the chords in bar 6.

- *Moulding Rule 2a:* If Moulding Rule 2 is applied to the top notes of the chords, then the result is played by Instrument 2 (Figure 7.12).
- *Moulding Rule 2b:* If Moulding Rule 2 is applied to the bottom notes of the chords, then the result is transposed down by one octave and played by Instrument 6 (Figure 7.12).

Figure 7.12 Moulding Rules 2a and 2b applied to bar 6.

- *Moulding Rule 3:* All descending sequences (see Rule 6) of notes are slurred in order to form an articulatory unit (Figure 7.13).

Figure 7.13 Moulding Rule 3 applied to bar 3.

- *Moulding Rule 3a:* If Moulding Rule 3 is applied to the top notes of the chords then the result is played by Instrument 1 (Figure 7.14).
- *Moulding Rule 3b:* If Moulding Rule 3 is applied to the top notes of the chords and if the result follows an outcome from the application of Moulding Rule 1, then the result is played by Instrument 2.

Figure 7.14 Moulding Rules
3 and 3a applied to bar 3.

• *Moulding Rule 3c:* If Moulding Rule 3 is applied to the bottom notes of the chords, then the result is played by Instrument 5 (Figure 7.15).

Figure 7.15 Moulding Rule
3c applied to bar 3.

• *Moulding Rule 3d:* If Moulding Rule 3 is applied to the bottom notes of the chords and if the result follows an outcome from the application of Moulding Rule 1, then the result is transposed down by one octave and played by Instrument 6 (Figure 7.16).

Figure 7.16 Moulding Rules
3c and 3d applied to bottom
notes of bars 3 and 4.

• *Moulding Rule 4:* Single top notes are by default played by Instrument 1 (Figure 7.17).
• *Moulding Rule 5:* Single bottom notes are by default played by Instrument 6 (Figure 7.17).
• *Moulding Rule 6:* A descending sequence terminates when the next note is ascending (Figure 7.18, top).
• *Moulding Rule 7:* An ascending sequence terminates when the next note is descending (Figure 7.18, bottom).

Figure 7.17 Moulding Rules 4 and 5 applied to bar 6. Note, however, that this bar is out of context here as these notes might form articulations with the next bar.

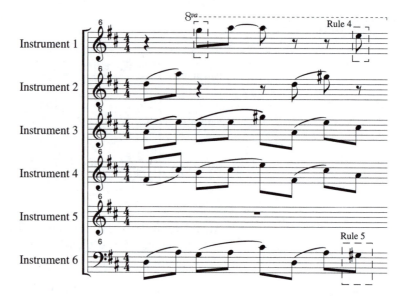

Figure 7.18 Moulding Rules 6 and 7 applied to notes of bars 6 and 4 respectively.

- *Moulding Rule 8:* If a descending sequence is followed by an ascending sequence or if a descending note is the starting point for an ascending sequence, then both sequences are merged and slurred in order to form a larger articulation and the result is played by the instrument of the former sequence (Figure 7.19).

Figure 7.19 Moulding Rule 8 applied to notes of bar 5.

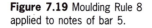

The result of the application of the moulding rules to the material of Figure 7.8 is given in Appendix 2. Bear in mind that the above rules are by no means exhaustive; for example, there are no rules for dynamics or for resolving conflicting situations. Also, it does not account for the idiosyncrasies of particular instruments as the instrumentation is not explicitly defined.

The three generative combinatorial sub-modules introduced above are generic in the sense that they could be applied to any number of chords of any nature, and not necessarily in the context of a tonal key. Also, the moulding rules are fairly generic apart from the fact that they are tailored for six instruments. More rules could be specified for enhancing the clip; e.g., a rule to apply a crescendo (in dynamics) to all ascending sequences, a rule to stress single notes, and so forth.

7.2 From form to content

This case study presents a technique for generating sequences of note strings using a simple *formal grammar* coupled with *distribution functions*; an introduction to formal grammars and distribution functions can be found in Chapters 2 and 3, respectively. As with the previous case study, the technique will be presented by means of a practical example, inspired by the work of Douglas Hofstatder (1979).

The grammar for our example is very simple: it comprises four production rules for generating strings from an alphabet of three symbols: ◆↑ and ↓. Given an initial token ◆↑, the production rules of the grammar are defined as follows:

- *Production Rule 1*: If the last symbol is ↑, then ↓ can be added at the end of the string.
- *Production Rule 2*: The string ◆* can be replaced by the string ◆**.
- *Production Rule 3*: If ↑↑↑ occurs in a string, then ↑↑↑ can be replaced by ↓
- *Production Rule 4*: If ↓↓ occurs in a string, then ↓↓ can be dropped
- *Addendum*: the symbol * stands for a variable that can hold both a symbol of the alphabet or a string.

An example of a sequence of eight strings resulting from the iterative application of the production rules to an initial token ◆↑ is given as follows:

◆↑	(initial token)
◆↑↑	(by applying Generative Rule 2 to the initial token)
◆↑↑↑	(by applying Generative Rule 2 to the previous result)

◆↓↑   (by applying Generative Rule 3 to the previous result)

◆↓↑↓↑   (by applying Generative Rule 2 to the previous result)

◆↓↑↓↑↓   (by applying Generative Rule 1 to the previous result)

◆↓↑↓↑↓↓↑↓↑↓ (by applying Generative Rule 2 to the previous result)

◆↓↑↓↑↑↓↑↓   (by applying Generative Rule 4 to the previous result)

The strings generated by the above grammar are regarded as abstract melodic forms in the sense that given an initial note, the symbols ↑ and ↓ correspond to the direction of the next note in the melody. The symbols of the alphabet are thus interpreted according to the following convention: ◆ represents a root note, ↑ indicates that the next note goes upwards in relation to the current one, and ↓ indicates that the next note goes downwards.

The actual interval, that is, the distance that separates two notes, is selected from a set of prescribed intervals according to a chosen distribution function. The distribution function dictates the probability that the elements of a set will be picked, in order to compute the pitch of the next note. In order to illustrate this, consider the set I of intervals as follows:

$$I = \{ 10, 20, 21, 30, 32, 45, 46, 55, 56, 61, 70, 71, 85, 86 \}$$

Each interval is represented by a code of two digits where the first digit indicates the *size* of the interval and the second indicates the *type* of interval; for example, 31 means an interval of a major third. The sizes are codified as:

1 = unison
2 = second
3 = third
4 = fourth
5 = fifth
6 = sixth
7 = seventh
8 = octave

The intervals can be of the following types:

0 = minor
1 = major
5 = perfect
4 = diminished
6 = augmented

Two examples of note sequences corresponding to the strings ◆↓↑ ◆↑↑↓↑↓ using *exponential distribution,* and to the string ◆↑↑↓↓↑↓↓↑↓ using *linear distribution,* are illustrated in Figures 7.20 and 7.21, respectively (not necessarily generated by the grammar above).

Figure 7.20 Note sequence using exponential distribution.

Figure 7.21 Note sequence using linear distribution.

As with the previous case study, the compositional method presented here deals only with note intervals, but it is perfectly feasible to extend it in order to deal with other musical aspects. All the same, composers may prefer not to tie all compositional aspects to the same kind of generative process; this procedure would not give much scope for intervention during the compositional process. Moreover, it is often more effective and interesting to explore different generative processes, one for each aspect of the piece; for example, a formal grammar for pitch, a Markov chain for rhythm, a cellular automata for form, and so on.

7.3 Phonetic grounding

The interplay between music and language is a recurring theme in this book as the author is a strong believer that both capacities share a common pool of cognitive strategies. The relationship between the sonority of the human voice and its potential to express emotions and ideas, not necessarily within the context of a specific language, is fascinating. People who have visited a foreign country with no knowledge of its language, often acknowledge that they could understand a limited repertoire of utterances independent of the spoken language. Such understanding emerges mostly due to the action of our musical cognitive machinery; there is very little that the brain can actually do without being familiar with the lexicon and grammar of the language in question.

Building a grammar-based system for generating music is one way to practically explore the kinship between language and music; grammars are introduced in Chapters 2 and 3. Another approach is to consider that phonemes and musical notes can be characterised by the same attributes: pitch, duration, dynamics and timbre. Musical systems can therefore be defined based upon the phonological properties of languages. Complementing the two previous case studies where grammars, probabilities and rules were employed to define prescriptions for generating musical passages, the following paragraphs illustrate how a note system can be defined based upon the timbre of phonemes.

7.3.1 Formants and timbre

The spectral contour of most vocal sounds has the appearance of a pattern of 'hills and valleys' technically called *formants* (Figure 7.22). The centre frequencies of the first three formants (represented as F_1, F_2 and F_3 in Figure 7.22) and their respective amplitudes are largely responsible for the characterisation of the *phonetic timbre* of a vocal sound. The notion of phonetic timbre is slightly different from the notion of musical timbre. Whereas the latter is normally associated with the identity of musical instruments or with the effect produced by combining different instruments, the former characterises the phonemes of a particular language, most notably its vowels. In essence both notions refer to the effect of resonating bodies (or chambers) on the spectrum of an acoustic signal, as introduced by Helmholtz in the nineteenth century (Helmholtz, 1954).

Figure 7.22 The spectrum of most vocal sounds has the appearance of a pattern of 'hills and valleys' technically called formants.

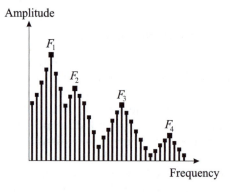

7.3.2 Understanding the vocal mechanism

Most of the time when we sing or speak, an air-stream is forced upwards through the trachea from the lungs. At its upper end, the trachea enters the larynx, which in turn opens out into the pharynx. At the base of the larynx, the vocal cords are folded inwards from each side, leaving a variable tension and a slit-like separation, both controlled by muscles in the larynx. In normal

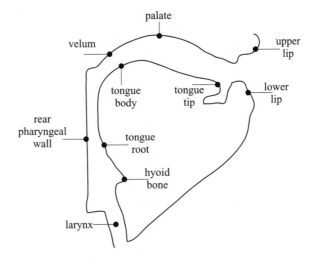

Figure 7.23 A model of the vocal tract featuring some of the most important components for synthesis control.

breathing, the folds are held apart to permit the free flow of air. In singing or speaking, the folds are brought close together and tensed. The forcing of the airstream through the vocal cords in this state sets them into vibration. As a result, the airflow is modulated at the vibration frequency of the vocal cords; the pitch of the sound is determined by this motion.

On its journey through the vocal tract, however, the spectrum of the sound produced by the vocal cords is altered by the resonance properties of the vocal tract. The vocal tract can be thought of as a cylindrical tube of approximately 16 cm long on average, closed at the larynx and open at the lips (Figure 7.24). Detailed accounts of the resonance properties of such a tube can be found in most books on acoustics (Howard and Angus, 1996).

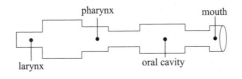

Figure 7.24 The vocal tract can be thought of as an irregular cylindrical tube closed at one end. Different sections of the tube are usually simulated by means of a series of filters.

Briefly, the resonant characteristic of a cylindrical tube is determined by the *standing waves* inside it. Roughly speaking, formants originate due to the action of such *standing waves*: those partials of the input signal that match the frequencies of the standing waves will pass smoothly (i.e., they will resonate) whereas all others will be attenuated. It is evident that the vocal tract is far from being a perfect geometric cylinder, but it is didactically plausible to think that the first formant F_1 is associated to the vocal tract's first standing wave with a pressure *anti-node* at the larynx and with a *node* at the lips. Similarly, the second format F_2 is associated to the second standing wave with a pressure anti-node in the region just above the middle of the tongue. The other formants result from similar principles.

7.3.3 Associating cause and effect

We can infer from the simple model described above that raising or lowering the tongue alone has relatively little effect on the value of the first formant because the body of the tongue lies between the anti-node and the node of the first standing wave in the vocal tract tube. Lowering the jaw, however, raises the frequency of the first standing wave as the jaw changes the diameter of the vocal tract in the region where there is a pressure node. The movement of the jaw is, therefore, one of the major determinants of the first formant. Conversely, the middle part of the body of the tongue lies where the second standing wave has a pressure anti-node. By the law of acoustics, if a constriction in a tube coincides with a pressure anti-node, then the frequency of the respective standing wave will increase. The tongue therefore is one of the major determinants of the second formant.

For instance, the vowel /i/ (as in the word 'it' in English) is produced by placing the jaw in a high position and by opening the lips only slightly; this results in a low first formant value of approximately 262 Hz. As the jaw is lowered and the mouth is opened towards the production of the vowel /a/ (as in the word 'car' in English), the first formant raises considerably. Then, as the jaw is raised towards the production of the vowel /u/ (as in the word 'rule' in English), the value of the first formant drops again. As for the value of the second formant, the tongue is more or less curved for the production of the vowel /i/, with its central part approaching the velum at the back of the roof of the mouth. This causes a constriction in the region of the pressure anti-node, which raises the value of the second formant to approximately 1800 Hz. The opening of the jaw towards the production of the vowel /u/, will also lower the tongue which consequently lowers the value of the second formant considerably.

To summarise, different vowels are associated with specific formant configurations and by changing the shape of our vocal tract we change these formants as we speak or sing. We could think of our vocal tract as a musical instrument that changes shape while it is being played. The three lowest formants are the most important for the recognition of vowels. On average, the first formant varies between 250 Hz and 1 kHz, the second between 600 Hz and 2.5 kHz and the third between 1.7 kHz and 3.5 kHz.

7.3.4 Synthesising formants

Subtractive synthesis is a well-known synthesis technique for synthesising formants. In subtractive synthesis, each formant is associated with the response of a band-pass filter (BPF). In this case, a composition of BPFs set to different responses is needed to synthesise a spectrum with distinct formants. This synthesis approach implies that the behaviour of the human vocal tract is determined by two main components: *source* and *resonator*, where the former produces a raw signal that is shaped by the latter (Figure 7.25). In order to simulate this phenomenon, a sound source of a rich spectrum is applied to the filter set, which in turn resonates those components of the spectrum that are compatible with the response of the filters, and attenuates the others. A subtractive formant synthesiser normally employs two types of sound sources: the *voicing source*, which produces a quasi-periodic vibration, and the *noise source*, which produces turbulence. The former generates a pulse stream intended to simulate the semi-periodic vibration of the vocal folds, whereas the latter is intended to simulate an airflow past a constriction or past a relatively wide separation of the vocal folds.

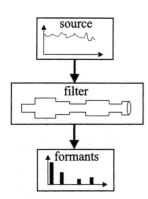

Figure 7.25 The source/resonator model.

In some circumstances, it might be desirable to control such a synthesiser in terms of the mechanics of the vocal system rather than in terms of filter parameters. In this case the synthesiser must be furnished with an interface that converts articulatory information such as the position of the tongue, jaw and lips into frequency, bandwidth and amplitude values for the filters (Figure 7.23). Utterances are then produced by programming the synthesiser to move from one set of articulatory positions to the next, similar to key-frame animation where the animator creates key frames and the intermediate pictures are automatically generated by interpolation.

7.3.5 Classifying phonetic timbres

Phoneticians have forged a framework to classify phonetic timbres (or vowels) in a three-dimensional space: *height*, *back-to-front position* and *roundness*. The first two axes are related to the position of the body of the tongue in the mouth and the third one is related to the shape of the lips. As with the previous example, this is an oversimplification of the vocal tract system, as there are more articulatory factors that determine the nature of the vocal sounds, such as the position of the jaw and the position of the larynx. Nevertheless, from an acoustic point of view, this model is capable of classifying the vowels of almost every language in the world. Figure 7.26 displays a set of reference sounds, known as the *cardinal vowels*, in the above cited three-dimensional space (Jones, 1956). Ladefoged and Maddieson (1996) proposed that five categories for height, three for back-to-front position and four for rounding are enough for categorising vowels, which amounts to 60 different possible classes (5 × 3 × 4). Some combinations are, however, very unlikely to occur: for example, it is most improbable that any language will hold the distinction of four degrees of rounding between low front vowels.

Figure 7.26 The cardinal vowels in the three-dimensional vowel space.

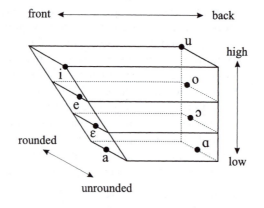

7.3.6 Towards grounded musical systems

Synthesis models and classificatory frameworks, such as the ones suggested above, are attractive to composers because they aid the definition of grounded musical systems. By grounded musical systems we mean systems in which their parameters are not entirely arbitrary, but rather are tightly coupled with realistic physical models, such as the vocal tract mechanism. As an example, let us define a pitch system based upon the formant structure of five different vowels, namely /a/, /e/, /i/, /o/ and /u/, as sung by a male and a female voice (total of 10 vowels): /a/ as in the word 'car' in English, /e/ as in 'bed', /i/ as in 'it', /o/ as in 'hot' and /u/ as in 'rule'.

First of all, we need to find the formant values for these vowels. One way of finding them is to analyse their recordings by means of a suitable spectral analyser (Figure 7.27). Even by looking at the visual representation of the analysis one should be able to infer the frequency, bandwidth and amplitude values of the lowest three formants; the program Audio Sculpt, developed by

IRCAM in Paris (the same producer as for OpenMusic, discussed in Chapter 8), provides good facilities for this. Once these values have been inferred, it is always good practice to feed them into a formant synthesiser in order to verify whether they are correct and to make the necessary adjustments.

Figure 7.27 The sonogram (a type of spectral analysis) of a vocal sound where one can clearly identify the formants of the signal.

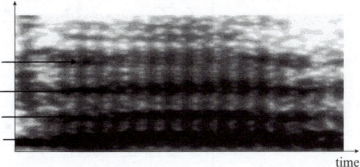

A chart containing the formant values for the ten vowels in question is given in Appendix 3. These values may of course differ slightly from person to person and according to recording conditions. But on the whole, the vowels should fall within the same classificatory boundaries. Note in our chart that we have quantised the frequency values to the nearest pitch of the standard 12-tone equal temperament system, with a tuning reference of A4 = 440 Hz. This quantisation is a compromise to render the data useful for composers working with standard musical notes. Although the quantisation should rarely alter the class of the respective vowels, composers working in the realm of synthesis and/or alternative tuning systems may not wish to quantise the formant frequencies.

Figure 7.28 portrays the result of our analysis in terms of standard musical notes. An additional fourth stave, labelled F_0, containing notes one octave below F_1, has been added at the bottom. These notes correspond to idealised natural fundamental frequencies of these vowels.

Having defined a note system such as the one in Figure 7.28, we could now couple it with a generative process for producing musical passages. For instance, it is perfectly feasible to adapt the combinatorial module introduced earlier in order to generate sequences of chords, and so on. An even more interesting approach here would be to devise a generative method based upon phonological properties and/or grammatical rules of specific languages.

Figure 7.28 A note system derived from formants.

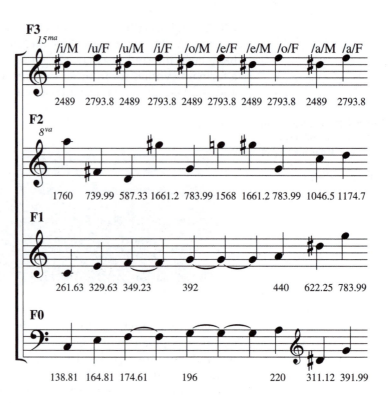

7.4 Final remarks

The three case studies presented in this chapter stand at the note level of abstraction (refer to Chapter 1), despite the fact that in the third case the notes emerged from the processing of spectral data. It should, however, be noted that the compositional processes given in this book are not restricted to generating notes alone, but that they also could be used to create higher level musical structures and lower level sound synthesis parameters. For instance, the idea of generating melodic forms using a grammar could well be applied to the generation of spectral envelopes. Another possibility would be to analyse the morphology of spoken utterances in order to build a framework to compose with sampled sounds at the building-block level.

The only limit is your imagination...

8 Music composition software on the accompanying CD-ROM

8.1 A programming language for algorithmic composition: Nyquist

Nyquist, developed by Roger Dannenberg at the Carnegie Mellon University, USA, is a programming language for music composition and sound synthesis: it supports both high-level compositional tasks and low-level signal processing within a single integrated environment. This feature is particularly attractive for composers wishing to work at the microscopic level of abstraction. Most sound synthesis languages (e.g., the popular Csound) are not entirely suitable for algorithmic composition because they are limited to instrument design: the actual notes to be played on an instrument must be manually specified one-by-one in a score file. There is no provision for writing generative musical programs in such languages.

Nyquist is implemented on top of a programming language called Lisp, more specifically XLisp, an extension of Lisp that is freely available for non-commercial purposes. Nyquist can be thought of as a library of Lisp functions that can be called up by other user-specified functions in a program. In fact, XLisp has been extended to incorporate sounds as a built-in data type, which gives Nyquist a great deal of expressive power. Composers thus write their algorithmic programs in Nyquist as if they were writing Lisp programs. The great advantage of this is that composers have the full potential of Lisp, which is a powerful language used in artificial intelligence research, combined with a wide variety of functions and abstract structures for composition devised by Dannenberg. The caveat of Nyquist is that composers should be familiar with Lisp programming in order to take full advantage of it. A brief introduction to Lisp is given in Appendix 4, but readers wishing to use Nyquist more effectively are encouraged to study further.

Helpful documentation for both Nyquist and XLisp is given in the Nyquist Reference Manual in the folder **nyquist**. Also, good introductory articles and a short tutorial can be found in *Computer Music Journal*, **21**(3), published in 1997.

Due to its Lisp background, Nyquist works interactively with the user via the Lisp interpreter's prompt. The symbol '>' means that the interpreter is waiting for your command. By typing the following line at the prompt we can readily make and play a sinewave:

> (play (osc 60 2.5))

The *osc* function activates a table-lookup oscillator which in this case contains a sinusoid. The first parameter 60 designates middle C as the pitch of the sound; it uses the same pitch number representation used by MIDI systems. The second parameter determines the duration of the sound: two and a half seconds. The result of *osc* is then passed to the function *play* which is in charge of generating the sound samples. Under Windows, Nyquist writes the samples in a sound file and plays it back automatically; under other systems the user might have to use a playback application to play the files. In the simple example given above, note that the sound will inevitably begin and end scrappily because there is no envelope controlling its amplitude.

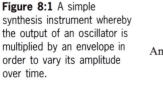

Figure 8:1 A simple synthesis instrument whereby the output of an oscillator is multiplied by an envelope in order to vary its amplitude over time.

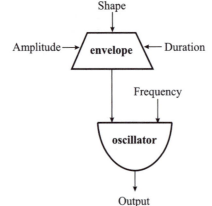

The following function defines a very simple instrument composed of an oscillator and an amplitude envelope:

```
> (defun my-instrument (pitch duration)
    (mult (env 0.05 0.1 0.5 1.0 0.5 0.4)
        (osc pitch duration)))
```

The function *env* generates a line-segment, which in turn is multiplied (function *mult*) by the output of the oscillator in order to vary the amplitude of the sound over time. We can now play this instrument by typing:

```
> (play (my-instrument c4 2.5))
```

Note that the above example uses a different notation for the pitch parameter. This is another way of saying middle C, where the number '4' indicates the octave of the note 'c'. Also, the output from *my-instrument* is much smoother than the output of previous *osc* example due to the action of the envelope.

A simple program to make a three-note chord on *my-instrument* could be defined as follows:

```
> (defun make-chord (duration)
    (sim (scale 0.3 (my-instrument c4 duration))
        (scale 0.3 (my-instrument e4 duration))
        (scale 0.3 (my-instrument g4 duration))))
```

The following line activates the *make-chord* program:

```
> (play (make-chord 3.5))
```

The *make-chord* program has two new pieces of information in it: *scale* and *sim*. The former is used to scale down the amplitudes of each note by 70% and *sim* is a construct that tells Nyquist to produce the three events simultaneously. In a similar way to *sim*, Nyquist provides another important construct called *seq*, which tells Nyquist to play events in sequence:

```
> (defun make-sequence (duration)
    (seq (make-chord duration)
        (make-chord duration)
        (make-chord duration)))
```

The above program uses *seq* to play a sequence of three chords, as defined by *make-chord*. Since there is no provision for inputting notes in *make-chord*, the *make-sequence* will play the same chord three times.

We are now in a position to illustrate one of the most powerful features of Nyquist: the ability to apply transformations to functions as if these transformations were additional parameters.

There are a number of transformations available in Nyquist and they are intelligent in the sense that they were programmed with default values that often perform their tasks without the need for major adjustments. As an example of this, suppose we wish to transpose the chord sequence produced by *make-sequence* five semitones higher. In order to do this we can simply apply the transformation *transpose* to the *make-sequence* function as follows (the parameter '5' after *transpose* indicates the number of semitones for the transposition):

```
> (play (transpose 5 (make-sequence 3.5)))
```

And indeed, transformations can be applied to functions within function definitions. Example:

```
> (defun two-sequences (duration)
    (seq (transpose 5 (make-sequence duration))
        (transpose 7 (make-sequence duration))))
```

A variation on the example above is given as follows:

```
> (defun two-sequences (duration)
    (seq (transpose 5 (make-sequence duration))
        (transpose 7 (stretch 4 (make-sequence duration)))))
```

In this case, the third line specifies a nested application of transformations where *stretch* is used to enlarge the duration of the chords in the sequence and then *transpose* is used to transpose the whole sequence seven semitones upwards.

Although not shown in the examples above, Nyquist can read and write MIDI files. MIDI files are loaded into a data structure and Nyquist can modify, add or remove MIDI messages from it. It is also possible to use Nyquist as a synthesiser to play back MIDI files. More information can be found in the reference manual and in the tutorial available on the CD-ROM. Nyquist runs under various operational systems, including Windows (95, 98, NT), MacOS and Linux/Unix. Two versions are provided on the CD-ROM: Windows and MacOS; the Linux version can be downloaded from the Carnegie Mellon University's web site. See software documentation for installation instructions, web site addresses and other information.

8.2 Visual programming: OpenMusic

OpenMusic, developed by Gérard Assayag and Carlos Agon at IRCAM in France, is a system for music programming that stands somewhere between a programming language and general composition software. In essence, it is a visual interface for the Lisp language that comes with a variety of built-in subroutines and abstract structures represented by icons. As with Nyquist, OpenMusic runs on top of Lisp, more precisely, the Macintosh Common Lisp (MCL). Composers construct programs in OpenMusic by dragging these icons onto a patching window and linking their respective inlets and outlets by wires (Figure 8.2). Behind the scenes, the pillar supporting this approach to visual programming is a paradigm favoured by the MCL's CLOS (Common Lisp Object System) known as Object Oriented programming. As OpenMusic's visual interface also incorporates CLOS, new methods and classes for implementing musical structures can be defined visually. Note that although MCL and Nyquist's XLISP share a similar fundamental syntax, both are different implementations of the Lisp language.

Figure 8.2 A program is built by connecting modules and patches in the patch window.

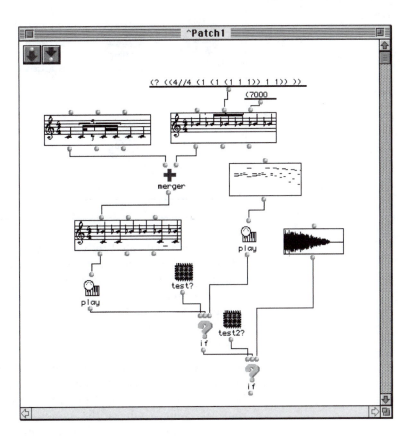

In OpenMusic jargon, the fundamental programming building blocks are called *modules*. There are various types of modules ranging from mathematical and logical operators, to musical generators and graphical editors. These are the Lisp functions provided by the OpenMusic system and they are represented by icons that can be dragged and dropped either from or to the *OM workspace* or the Macintosh finder. User-made programs are called *patches*, and patches can be embedded into other patches. The OM workspace operates very similarly to the Macintosh finder: it can hold folders containing patches and folders containing sub-folders and so on (Figure 8.3). A special folder that appears in the OM workspace when OpenMusic is loaded is the *Packages* folder; this folder contains the modules for making patches.

Figure 8.3 The OM workspace operates in a very similar way to the Macintosh finder.

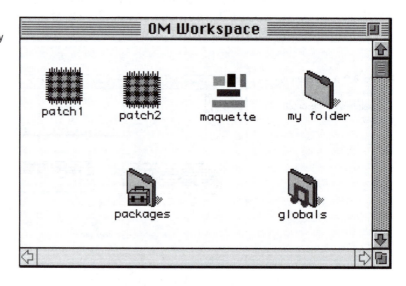

A new patch is created by selecting the option *New Patch* in the File menu. A new patch icon will appear in the OM workspace and the patch window can be opened by double-clicking on the icon. In order to build a patch in the patch window one drags modules (e.g., from the packages folder) onto the patch window and interconnects them; Figure 8.2 illustrates an example of a patch.

As far as musical form is concerned, OpenMusic features a type of patch called a *maquette*. A maquette is a special patch for creating hierarchical musical structures in time. It works as a kind of musical canvas where the composer can include modules, patches, sound files, MIDI files and other maquettes (Figure 8.4). Building blocks in a maquette can be incorporated in a functional way. Upon evaluation, the musical contents of the blocks are computed in an order that depends on the topology of the interconnection network. Then the maquette can be

Figure 8.4 The maquette is a special patch with a time dimension on the horizontal axis.

performed, normally from left to right. As blocks can be displayed graphically after they are computed (e.g., music notation, piano-roll, sound signal) the maquette could well be considered as an extended score.

The OpenMusic kernel is now available through GPL (Gnu Public Licence). A copy of OpenMusic (for Macintosh platforms) including sources, user documentation, developer's documentation and tutorial patches can be found in the folder **openmusic** on the Macintosh section of the accompanying CD-ROM.

8.3 Intelligent riff blocks: Music Sketcher

Music Sketcher is a prototype program that embodies some of the technology for music processing that is being developed at the IBM Computer Music Center in the USA (Abrams *et al.*, 1999). Although the authors themselves suggest that Music Sketcher should not yet be considered as a fully fledged composition program, it is already capable of producing very interesting results.

Music Sketcher is designed for composing at the building-block level, as discussed in Chapter 1: that is, the building blocks of

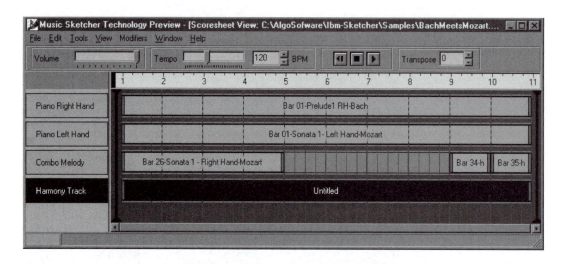

Figure 8.5 The main window of Music Sketcher can be thought of as a canvas where the composition is put together.

the composition are not single musical notes here but prefabricated short musical sections called riff blocks. The system provides a library of riff blocks where the composer can select and drag riffs onto a score (Figure 8.5). As with standard musical scores, the Music Sketcher score is divided into parts (or voices) and the riff blocks appear on these parts as rectangular boxes. The composer sees a composition as a sequence of rectangular boxes distributed between the parts. Once a block is on the score, it can be stretched by dragging its right-hand edge; this procedure will repeat the riff a number of times.

The creative potential of Music Sketcher resides in its ability to apply transformations to riff blocks. These transformations, or 'modifiers' in Music Sketcher's parlance, are graphical curves that can be applied to various musical aspects of music content, such as pitch, articulation or loudness. A 'velocity' modifier could be applied in order to cause a crescendo or a 'duration' modifier could be applied to halve the duration of the notes, to cite but two examples. Modifiers can also be applied to a block, a part or to the whole piece thereby allowing the reshaping of musical aspects at any structural level, and their combined effect can range from subtle expressive nuances to significant reworking of compositional materials. Using a modifier is fairly straightforward: first the user selects the aspect of the riff that he wishes to modify (e.g., pitch, duration, loudness, onset, etc.) and he draws a curve that describes the desired change over time. If more than one modifier is applied to the same musical aspect, then he can determine the order in which they are applied.

Another interesting feature of Music Sketcher is a built-in intelligent harmonic framework, called SmartHarmony (Abrams *et*

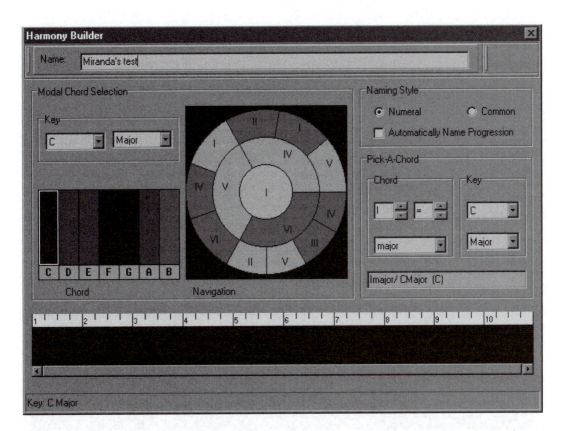

Figure 8.6 The Harmony Builder is a tool for building chord progression for the Harmony Track.

al., 2000). SmartHarmony embodies a model of tonal music that binds the music to a user-specified chord progression. This chord progression is put on the Harmony Track, situated in a special part at the bottom of the score (Figure 8.6). SmartHarmony then adjusts the riff's pitches so that they conform to the underlying chord progression. With this mechanism, the user can change the harmonic progression, reshape the melody, or even insert music from another context, while in all cases achieving a harmonically sensible result. The chord progression for the Harmony Track is specified via the Harmony Builder tool (Figure 8.8). This tool features a concentric circle on the left-side of the panel, which is a navigation guide that helps the composer create a sensible chord sequence within a given tonal key. The tonic chord is at the centre of the circle and the first circular row around it shows the chords that would normally follow the tonic chord. Then, for each chord in this row, the adjacent chords in the outer row are candidates for the subsequent elements of the progression.

Music Sketcher runs under Windows and a fully working version, as well as other informative material, can be found in the folder **sketcher**. The user manual and a useful tutorial are provided via the program's Help menu. Both references mentioned above are on the CD-ROM in PDF format, in the folder **sketcher\papers**.

8.4 Hybrid approach: Tangent

Tangent, developed by Paul Whalley, is a program that combines various generative and transformation techniques within a single composition environment. Through careful combination of these elements the system is capable of producing interesting and coherent musical output.

The program is based upon the top-down approach to composition discussed in Chapter 1, in the sense that it fosters the creation and manipulation of musical structures rather than individual notes. Basically, it works by creating sequences of musical passages (or 'periods' in Tangent terminology); the composer can specify different ways to generate any of them. In short, each period has ten parameters that can either be manually specified or automatically generated. Examples of

Figure 8.7 The main window of Tangent.

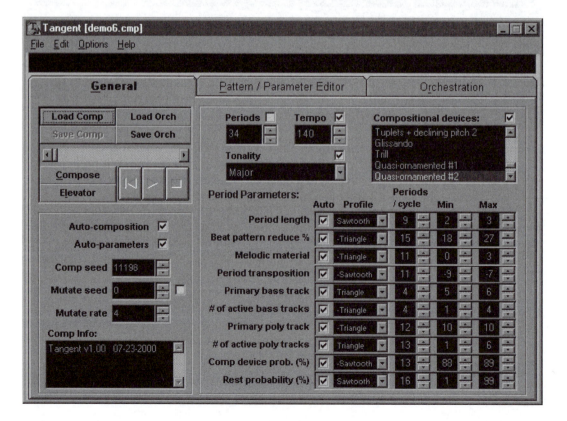

parameters are the length of the period, the source of melodic material and transposition (Figure 8.7).

Briefly, the compositional process in Tangent works as follows: it starts by producing a musical pattern that may be repeated a number of times, according to the length of the period. The subsequent period results from the application of some transformations to this pattern, and so on. In order to apply these transformations, Tangent calls up various pre-programmed *compositional devices*. For example, the *Dynamic contrast of pitch (+)* device will emphasise high pitch notes by increasing the MIDI velocity value of all notes that are higher than the average pitch of the notes of the respective period. There are over 20 devices available including rhythmic contrast of pitch, glissando, trill, random rests, and so on.

Tangent can produce up to 16 voices and each voice can be set to play the musical patterns in different ways, thus allowing for further characterisation of the individual voice; e.g., loudness, transposition, whether the notes are played legato, staccato, etc.

On the CD-ROM there is a demo version of Tangent for Windows in the folder **tangent**. The limitations of the demo do not prevent the reader from using it, but you cannot save the results in MIDI format. Readers are encouraged to register in order to support the developer; instructions for registering are given in the user manual. The documentation is well written and very easy to follow.

8.5 Internet: the SSEYO Koan system

The SSEYO Koan system comprises a number of programs for creating and playing back music. In fact the system is better seen as a kind of platform around which many SSEYO products are built. It has been designed by SSEYO, a British company founded by Tim and Peter Cole that is dedicated to the development of Internet-based music technology. The Koan system is therefore ideal for creating music for the Internet: a Koan-generated piece can be as small as 1 Kb. All that the visitor's browser needs to play a piece from a web site is a plug-in that is freely available from SSEYO.

The SSEYO Koan product range is based upon four main programs: *Koan X, Koan Pro, Koan File Player* and *Koan Plugin*. Whereas the first two are authoring tools, the last two are for playing back Koan pieces; Koan Plugin features some facilities for interactive web music making and can be used as the engine to power a diverse range of Internet audio applications.

Koan X works as a kind of drag'n'mix tool where the composer can choose various pre-composed musical templates from a menu and drag them onto a working area called the Mix Window (Figure 8.8). A menu of templates is provided where the user can chose drum patterns, ambient voices, arpeggios, bass lines, chords and so forth, but what is interesting here is that these templates are not fixed sound files or MIDI sequences. Rather, they are instructions for generating short musical passages: these can be fixed melodies, random note sequences, effects, etc. The templates can play either SoundFonts, Audio files (e.g., WAV and MP3) or MIDI, in addition to SSEYO's own SoftSynth, a software synthesiser. By positioning the icons of the templates in different areas of the Mix Window we can change the loudness or the stereo position of the respective output: the vertical axis corresponds to loudness and the horizontal axis corresponds to stereo position. There is also a facility whereby we can set the templates in continuous motion in the Mix Window, according to a number of given trajectories.

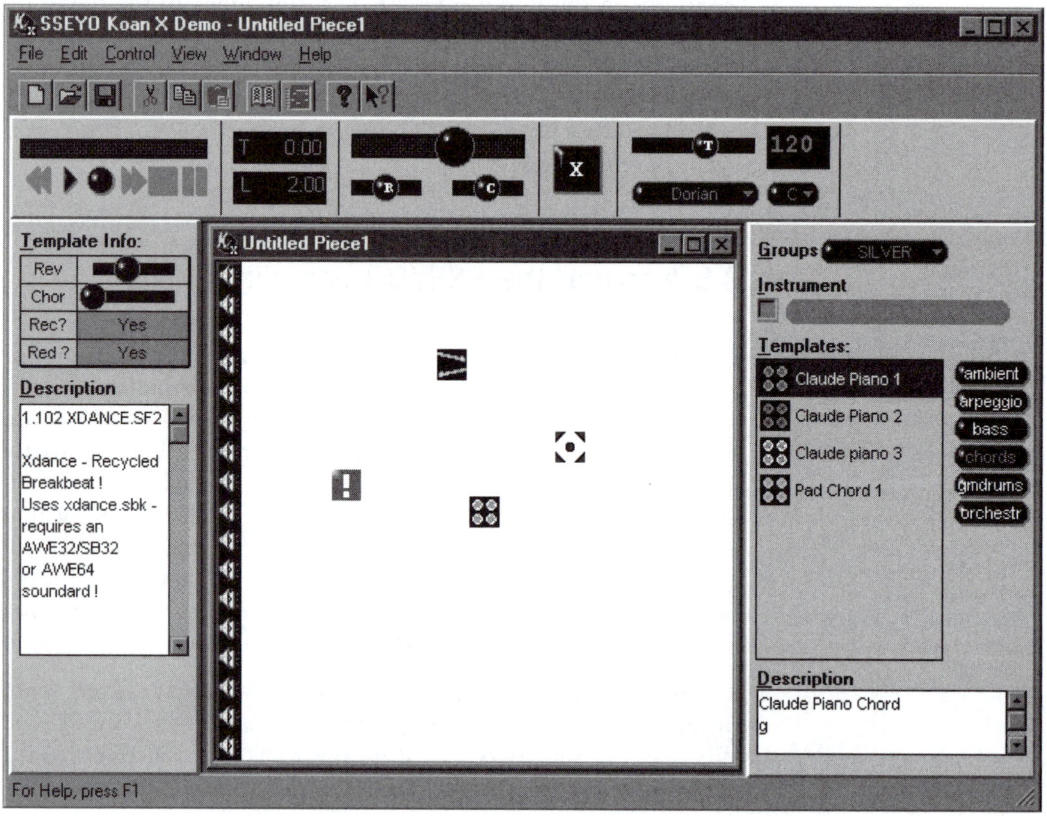

Figure 8.8 Musical templates are dragged onto the Mix Window of Koan X in order to produce music.

Koan Pro is a fully fledged piece of composition software for creating both templates for Koan X and entire pieces for use with the Koan Plugin, for example. Koan Pro provides a number of parameters for composition, some of which can be controlled by time-varying functions (or 'envelopes' in Koan jargon) throughout the entire piece (or template), whilst others can be fixed throughout the piece. Each piece contains a number of voices with a number of voice level parameters whose values will vary the way in which the music is generated. One of the most important factors determining which notes each voice will play are given by various rules, grouped as Scale rules, Harmony rules, Next Note rules and Rhythm rules. Each of these rules determines the probabilities of various events occurring. For instance, if the probability of a note C# occurring in a Scale rule is zero, there is no chance of C# occurring at all. The benefit of giving each event a probability means that it is possible to obtain many subtleties of note combinations. Koan Pro calculates which pitches to play from the Harmony and Scale rules and the durations from the Rhythm rules. If we set up and use a Scale rule to mimic a major scale (where only certain notes are possible), then we can be assured that only notes in that major scale will play, according to the probabilities we have assigned to the various intervals. Also, Koan can be used to augment generatively any track of a given MIDI file.

The SSEYO Koan Music Engine (SKME), is the real-time music generative engine under the hood of the Koan system. Whenever a Koan piece is played, the SKME interprets the parameter setting of the templates of the piece in order to generate the music in real-time. The SKME can give performances that are different each time, according to over 250 variable settings. Because the SKME's musical output can be different every time Koan provides an interesting element of originality and interest.

The Koan software runs on both Windows and Macintosh platforms. In the folder **koan** there are demo versions of Koan Pro and Koan X, plus Koan Plugin and Koan File Player (Koan X and File Player are not available for Macintosh).

8.6 Grammars and constraints: Bol Processor

Bernard Bel started the development of Bol Processor in the early 1980s while he was working in India but he continued to develop it when he joined the Artificial Intelligence Laboratory of the Centre National de la Recherche Scientifique (CNRS) in Marseille, France, in 1986. The newest version of Bol Processor, BP2, was produced a couple of years ago while Bel was working at the Centre for the Human Sciences in New Delhi, India. Srikumar Karaikudi Subramanian has recently joined the venture.

BP2 software is specially devised for composing using grammars (Bel, 1998). Originally, Bol Processor was designed for ethno-musicologist Jim Kippen, who was looking for a systematic way to study tabla drumming improvisation. North India's tabla drumming improvisation is bound by rules for determining the correct sequences and Kippen wanted to formalise grammars for tabla music. Tabla musicians often represent tabla sequences using onomatopoeic-like mnemonics that roughly correspond to different drum-strokes; e.g., dha, ti, ge, na, tirakita (trkt), dhee, tee, ta, ke, etc. (Kippen and Bel, 1992). These mnemonics are called bols, hence the name 'Bol Processor'.

The notion of *sound-objects* is the crux of BP2: sound-objects correspond to the terminal nodes of a grammar (see Chapter 2), and a sound-object may correspond to either a sequence of MIDI codes or sound synthesis commands. A single musical note is a sound-object with a single MIDI code for triggering it. Notes can be written in English, European or Indian notation in addition to a MIDI key number; for example 'A4', 'la3', 'dha4' and 'key#69' are equivalent. An example of a simple BP2 grammar whose terminal nodes are all single notes could be given as follows (the outcome from the above grammar is given in Figure 8.9):

S → A B B A
A → 'do3'
B → 'mi3' 'sol3' C
C → 'si4' A

Figure 8.9 An example of a musical passage generated by a grammar (see main text). Rhythmic aspects have been omitted for the sake of clarity.

Perhaps one of the most interesting features of BP2 is a mechanism to handle *polymetric expressions* (i.e., strings describing concurrent sequences). For example, consider a sequence of five sound-objects {a, b, c, d, e} running in parallel with another sequence of three sound-objects {f, g, h}. In this case, the polymetric algorithm will superimpose these two expressions by stretching them as follows:

a _ _ b _ _ c _ _ d _ _ e _ _
f _ _ _ _ g _ _ _ _ h _ _ _ _

The '_' symbol means that the preceding sound-object has been stretched in time; in this case a sound-object may have been either stretched according to a given tempo, or cycled a number of times (N.B. a sound-object can also be contracted). Sound-objects may be performed either with a regular beat (*striated time* in BP2's parlance) or with no beat at all (*smooth time*); the polymetric algorithm can also handle multi-layered sequences.

When the terminal nodes correspond to more complex sound-objects rather than single notes one can really begin to exploit the full potential of BP2. Moreover, due to its tabla music origins, BP2 is equipped to manage complex timing structures. Sound-objects can have their own embedded metrical and topological properties, such as whether the beginning sound-object may overlap the end of a preceding sound-object, or whether the end of a sound-object should be truncated if it overlaps the subsequent sound-object (Figure 8.10).

Figure 8.10 Bol Processor's Data window and Control Panel with a grammar and a glossary visible at the bottom.

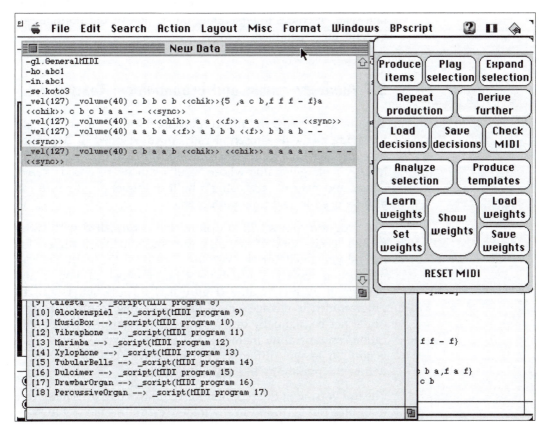

The idea of assigning properties to sound-objects comes from the notion of *constraint-satisfaction* programming developed by computer scientists to solve large combinatorial problems. The association of generative grammars with constraint-satisfaction programming makes BP2 a very powerful tool for defining musical grammars. Despite this, BP2 comes with a number of tools for enhancing the grammar such as context-sensitive substitutions, dynamic weighting of rules and serial music routines (see Chapter 2).

BP2 runs on Macintosh under MacOs 7.1 or higher and needs Opcode OMS for MIDI. BP2 is a shareware product. The full version of BP2 plus documentation and examples is provided on the CD-ROM (folder **bp2**) but readers wishing to use it regularly are required to register in order to support its developers (refer to file **registration.txt**). Users are also welcome to join an e-mail discussion group on the Internet; the address is given in the documentation. In addition to MIDI, BP2 also can generate Csound scores (Csound is programming language for sound synthesis; more information is given in the BP2 documentation) in order to synthesise music. This feature is particularly important when it comes to generating music that would normally be beyond the capabilities of MIDI.

8.7 Aleatory music and probabilities: Texture

Texture is a program for composing aleatory music (music involving elements chosen at random) developed in Argentina by composer Luis María Rojas. Basically, Texture is an aleatory MIDI events generator whose results can be recorded, saved, edited and played back using a built-in sequencer. Texture can also be used in real-time performance.

The program comes with a Control Panel furnished with slides for the specification of the minimum and maximum values that can be generated for the various aspects of the piece such as notes, loudness (i.e., MIDI velocity), number of voices and density of events in time (Figure 8.11). These sliders can be programmed to change in real-time according to functions that can either be manually drawn or created by any suitable application for generating mathematical curves. Texture also offers a facility to assign different probability weights to the values within the minimum and maximum intervals.

The full version of the program for Windows patforms and some examples can be found in the folder **texture** on the CD-ROM.

Figure 8.11 The Control Panel of Texture.

8.8 Number theory: MusiNum

MusiNum, designed by Lars Kindermann, is a program that generates MIDI notes from number sequences. The musical engine boils down to a counter whereby the user can specify different counting modes and the base system for the numbers. At each step the counter produces a MIDI note.

In order to understand how MusiNum produces the notes, let us take as an example a counter that counts binary numbers in single steps: 0, 1, 10, 11, 100, etc. In this case, the notes are computed by adding up the digits that are equal to one in each of these strings of binary numbers. If the result is equal to one, then it corresponds to the note C; if the result is equal to two, then it corresponds to the note D; if equal to three, to the note E; and so on (Table 8.1 and Figure 8.12).

A variety of different note sequences can be obtained by setting the counter with different step modes and base systems via the

Table 8.1

Decimal number	Binary counter	Amount of 1s	Musical note
1	1	1	C
2	10	1	C
3	11	2	D
4	100	1	C
5	101	2	D
6	110	2	D
7	111	3	E
8	1000	1	C
9	1001	2	D
10	1010	2	D
...	etc ...	...	...

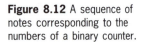
Figure 8.12 A sequence of notes corresponding to the numbers of a binary counter.

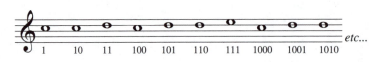

Voice Edit window (Figure 8.13). Also, different settings can be specified for different voices in order to generate polyphonic material.

MusiNum is simple to operate and yet it is an excellent tool for exploring the musical potential of self-similarity in number sequences (e.g., if you play every second note of the sequence in Figure 8.12, the result will be the same melody). The full version of MusiNum for Windows can be found on the accompanying CD, in the folder **musinum**.

8.9 Iterative algorithms: a Music Generator

Arnold Reinders' *a Music Generator* is a program that renders music from iterative algorithms such as fractals and chaos (Chapter 4); it also can render music from bit map images, text and numbers.

The program consists of two main sections: the data section and the MIDI section. The former contains the data from which the music generator will produce musical material (e.g., a fractal) whereas the latter holds instructions on how to play the music. Here the composer can define musical scales, control loudness and time, determine the length of the composition and so on.

On the top right of the main program window (Figure 8.14) there are several tabs, each of which contains a group of functions named: 1D Fractals, Dynamical Systems, Complex Maps, Data

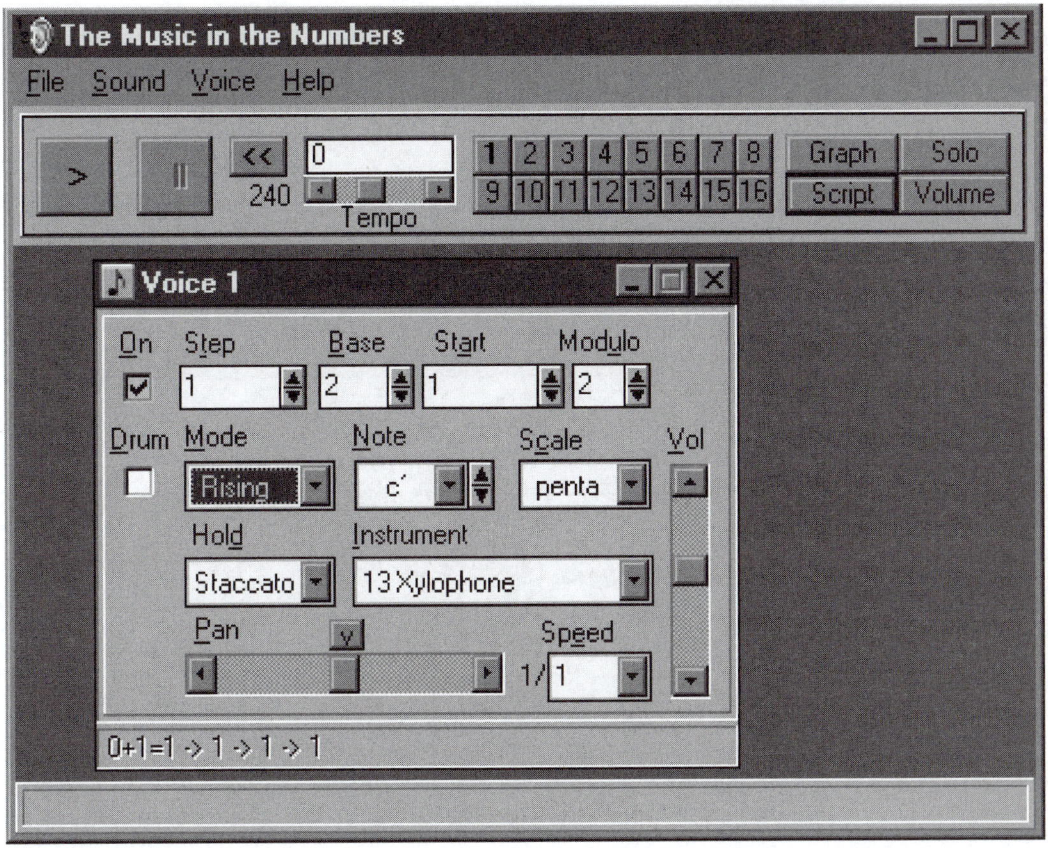

Figure 8.13 The Voice Edit window of MusiNum.

and Others. These functions produce the contents for the data section. A simple click on one of the icons representing the functions will place it in the data section on the left of the main window. Once a function has been placed in the data section the user can edit it in a number of ways and a plotting facility is provided in order to visualise its output.

In order to turn the data into music, the composer must allocate the functions on the data section to the controllable aspects of the MIDI section, on the right of the data section. There are four musical aspects that can be controlled by a function: Notes, Duration, MIDI Speed (that is, loudness) and Time; these correspond to the four buttons at the lower left side of the MIDI section. Functions are allocated by dragging its icon from the data section onto one of these buttons.

The musical results can be displayed on a built-in 'piano roll' while it is being generated, and then saved onto a MIDI file in case the user wishes to load them into another application. It

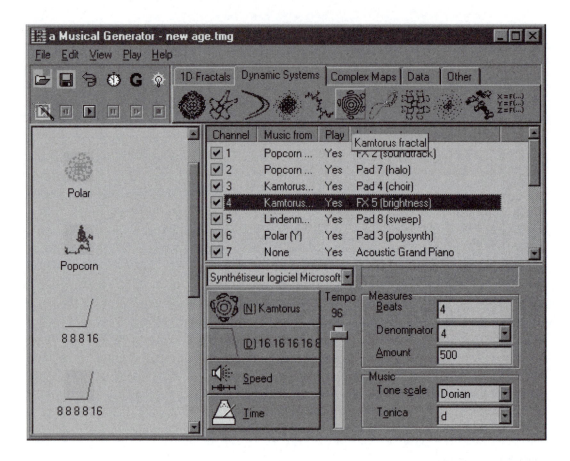

Figure 8.14 The main window of a Musical Generator.

also has the capacity to synchronise the program with other packages via the MIDI time clock.

A Music Generator is a shareware program for Windows platforms, manufactured and distributed by MuSoft Builders in The Netherlands. The full version is provided in the folder **musigen** but readers are kindly required to register after 30 days of use.

8.10 Fractals: FractMus

FractMus is a program that generates music from iterative algorithms, designed by Spanish pianist and composer Gustavo Diaz-Jerez. There are a variety of iterative algorithms available, some of which were discussed in Chapter 4: Morse-Thue, Logistic Map, 1/f Noise, Henon, Hopalong, Martin, Gingerbread man, Lorenz, 3n+1 numbers, 1D Cellular Automata, etc. These algorithms are fully explained in the Help documentation of the program.

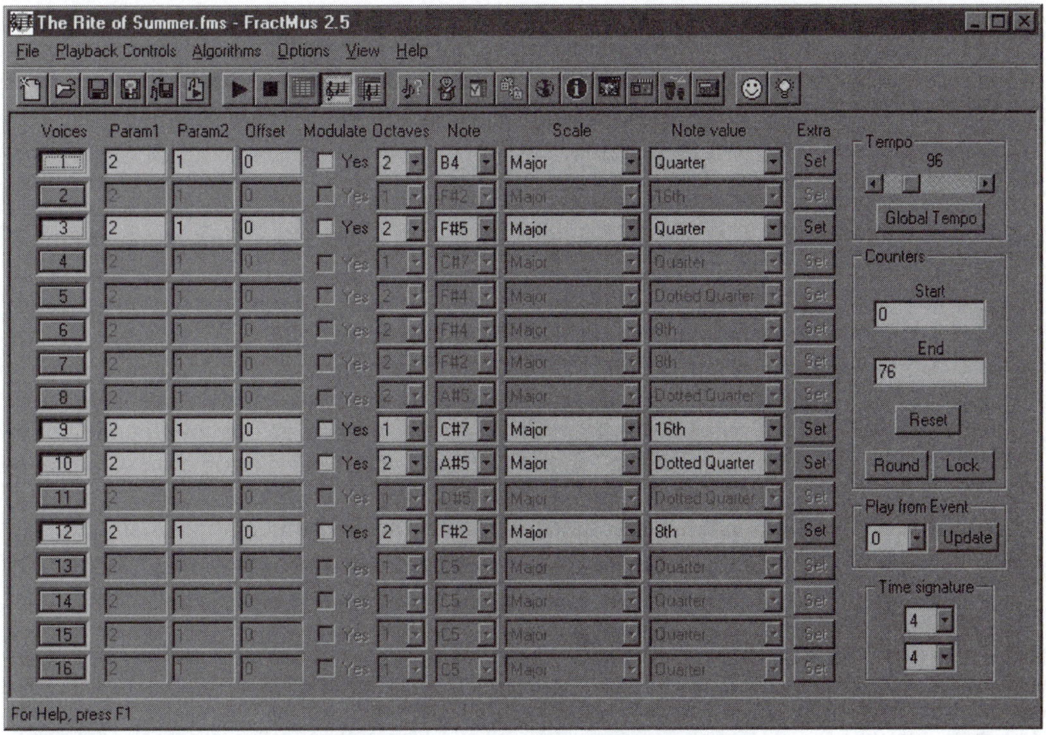

Figure 8.15 FracMus is capable of producing astonishingly good quality material from iterative algorithms.

FractMus is straightforward to use: there are up to 16 voices, each of which is allocated one iterative algorithm by the composer. In addition, the composer specifies a number of parameters such as a musical scale, an initial note and the basic rhythmic figure for each voice (Figure 8.15). The algorithms iterate according to a counter that monitors the beat of the different voices and at each iteration a new note is picked from a respective scale, according to the result of the iteration. FractMus is capable of producing astonishing musical material. The full version of FractMusic for Windows can be found in folder **fractmus** on the accompanying CD-ROM.

8.11 Cellular automata: CAMUS

CAMUS is a system for composition using cellular automata developed by the author and implemented for the Windows platform by Kenny McAlpine at the University of Glasgow in Scotland. There are two programs on the CD-ROM: the original *CAMUS*, as proposed by the author in an article published in *Interface/Journal of New Music Research*, Vol. 22, 1993, and *CAMUS 3D*, which is a variant proposed by McAlpine in collaboration with Stuart Hoggar.

The underlying concepts behind CAMUS were discussed in Chapter 6 and both programs are furnished with comprehensive documentation and tutorials in the folder **camus**. In short, both versions of the system render music from cellular automata by mapping live cells from the Game of Life to three-note (CAMUS) or four-note (CAMUS 3D) chords. In the latter case, for example, the x cell co-ordinate is used to describe the interval in semitones between the fundamental note of the chord and the lower internal note, the y cell co-ordinate is used to describe the interval in semitones between the lower internal note and the upper internal note, and the z cell co-ordinate is used to describe the semitone interval between the upper internal note and the top note of the chord. In order to establish which MIDI instrument will play a live cell, the system looks at the state of the corresponding cell in the Demon Cyclic Space automaton (Chapter 6, Figure 6.10). The instruments that play the music for each channel are specified using the Instrumentation dialogue box, where the user can associate a general MIDI instrument with each possible state of the Demon Cyclic Space automaton.

Figure 8.16 CAMUS 3D consults a Markov table in order to determine the order in which the notes of a cell will be played.

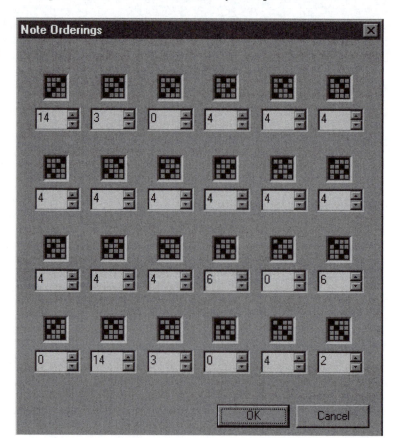

The internal timing structure of the cells (i.e., the order in which the notes of the chord will be played) is dealt with differently by CAMUS and CAMUS 3D. Whilst CAMUS computes this according to an intrinsic codification system (Chapter 6, Figure 6.11), CAMUS 3D consults a user-specified Markov table (Chapter 3). In this table, the temporal positions of each of the notes are represented by a 4×4 grid (Figure 8.16). The bottom row of boxes represents the lowest note of the chord, the second bottom row represents the lower internal note, the second top row represents the upper internal node and the top row represents the top node. The order of the shaded cells from left-to-right determines the order in which each of the notes is played.

8.12 Genetic algorithms: Vox Populi

Vox Populi is an interactive composition program that uses genetic algorithms (GA), developed by Jônatas Manzolli, Artemis Moroni, and co-workers at the Universtity of Campinas, Brazil. The program is for Windows platforms and it can be found on the CD-ROM in the folder **populi**.

In Vox Populi, GA are used to evolve a set of chords, or *population of chords* in GA jargon. Each chord of the population has four notes, each of which is represented as a 7-bit string. A chord is thus represented as a string of 28 bits (e.g., 1001011 0010011 0010110 00101010) and the genetic operations of crossover mutation are applied to this code in order to produce new generations of the population (Chapter 6). The fitness criteria take into account three factors: melodic fitness, harmonic fitness and voice range fitness. The melodic fitness is evaluated by comparing the notes of the chord to a user-specified reference value. In short, this reference value determines a sort of tonal centre, or attractor: the closer the notes are to this value, the higher the fitness value. The harmonic fitness takes into account the consonance of the chord and the voice range fitness measures whether or not the notes of the chord are within a user-specified range.

A straightforward user-interface provides sliders and other controls for the specification of fitness parameters and other musical attributes (Figure 8.17). Vox Populi can be used for real-time performance. The interface features a drawing area where the user can draw lines to control the reference values for fitness measurements and rhythm in real-time. This graphic control is interesting from a performance point of view as it suggests metaphorical conductor gestures as if the performer were conducting an orchestra. Manzolli himself composes music

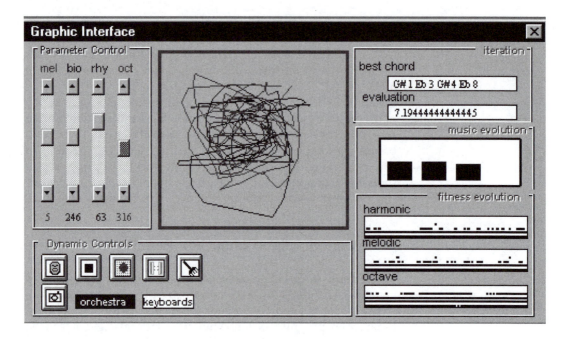

Figure 8.17 The Vox Populi main window.

using a glove equipped with sensors that translate the movements of his hands onto this graphic pad.

8.13 Selective automata: Harmony Seeker

Harmony Seeker combines the generative ability of cellular automata (CA) with the selective mechanisms of genetic algorithms (GA) in a unified environment for music composition. It was developed and implemented by Valerio Talarico, Eleonora Bilotta and Pietro Pantano, at the University of Calabria, Italy. Basically, Harmony Seeker renders the output of a population of one-dimensional CA into MIDI data, but since it is extremely unlikely that a newly created population of CA will produce a satisfactory outcome right at the outset, the system employs GA technology in order to improve the performance of the population.

The philosophy behind the system is as follows: a single CA is considered as an individual whose characteristics are determined by its transition rules (Chapter 6) and the initial states of the cells; note that the same set of transition rules will invariably produce a different outcome when applied to distinct initial cell configurations. The system starts with a population of CA, each of which contribute to the musical output and its objective is to improve the performance of this population by finding optimal rules and initial cell configurations. This is done by means of GA

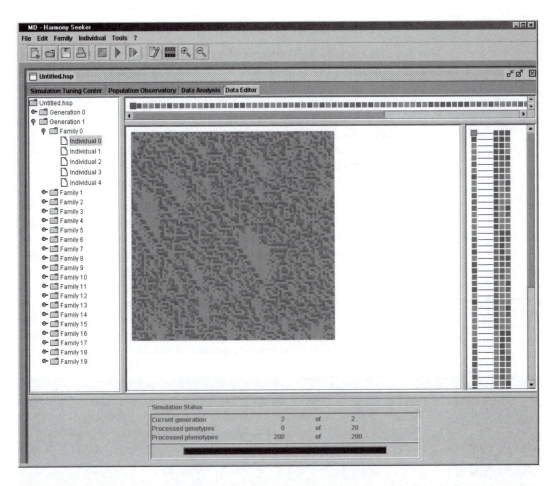

Figure 8.18 Harmony Seeker's main window.

techniques, such as mutation and crossover, whereby the selection criteria are based upon a consonance measurement function.

Harmony Seeker's main window features various tab panels for tuning the various parameters of the system and monitoring its performance. The main tab is *The simulation tuning centre* where the user selects the nature of the consonance measurement, methods to render MIDI data from the CA and types of genetic algorithms; each of these aspects has it own user-definable settings. The other tabs provide graphs for displaying the evolution of the population and means for intervening during the evolutionary process (Figure 8.18).

On the CD-ROM, in the folder **harmseek** in the Windows section, there is a fully working version of Harmony Seeker programmed entirely in Java. It functions on any platform running the Java Runtime Environment (JRE) 1.3 or later (JRE is provided with the program on the CD-ROM); it has been tested on Windows, Linux and Sun Solaris.

8.14 Brain interface: IBVA system

The Interactive Brainwaves Visual Analyser (IBVA) System is an affordable brain wave analyser developed by IBVA Technologies in the USA. The IBVA takes brain wave signals of up to 60 Hz from electrode sensors on the scalp and performs a number of operations with it. The raw information from the electrodes can be seen in the Raw Data window like a conventional electroencephalogram display, but the IBVA also performs a windowed Fast Fourier Transform (FFT) analysis on this signal in order to represent it in terms of a series of frequency and amplitude values (Figure 8.19).

The interesting aspect of the IBVA system is that the user can allocate actions to different frequency bandwidths of the analysis results. When specific frequencies reach a given threshold value, then the system triggers the actions that are associated with the bandwidth where the frequencies fall. The system targets the *long-term coherent waves* (Chapter 5) emanating from our brain activity, that is: the alpha, beta, delta and theta frequencies (ranging from 2 Hz to 45 Hz); it can also capture the low-frequency signals (below 2 Hz) connected with eye movements. It is therefore possible to associate the movement of

Figure 8.19 The main window of the IBVA software featuring an FFT analysis and the electroencephalogram-like plot.

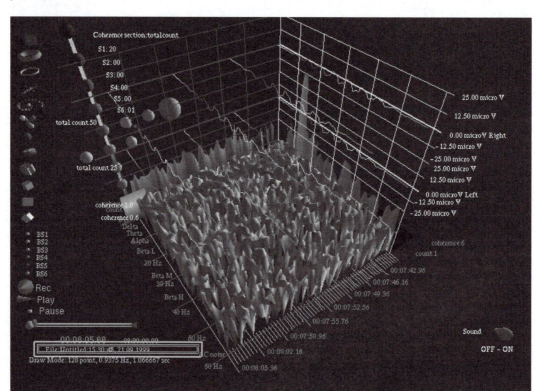

the eyes plus the four long-term coherent wave bands, with specific tasks, such as for example, controlling a MIDI synthesiser. The IBVA system comes with the Alps+ Orchestration Module which facilitates this specification. But Alps+ goes beyond simple MIDI note mapping, as it allows for the control of higher-level musical structures in a variety of ways.

The IBVA system comes with a head band containing electrodes, but it is possible to use other types of captors as well. The system is wireless; there is a small transmitter that can be attached to the head band with a receiver connected to the computer. This is a very interesting aspect from a musical point of view in case the composer wishes to use the system for live performance.

IBVA's MIDI control abilities offer endless possibilities to the composer, especially when used in conjunction with other systems for live performance. For example, Professor Sylvia Pengilly used the IBVA system coupled with the Max software, (commercialised by Cycling'74; see M below) to compose *Interface*. In this piece, pitches from brain wave signals are sent to a Max subroutine (or 'patch' in Max jargon) which basically echoes the incoming pitch in a series of ascending notes in perfect fourth intervals. This creates a basic texture of melodic contour due to the constant rise and fall of the incoming pitches, and to the overlapping of the perfect fourth notes. In order to create contrast and forge a sense of musical form, the Max patch was programmed in such a way that when the performer's brain activity reaches a specific level, a pre-composed sequence of percussive-like sounds is triggered. The end of this sequence triggers a clock device that allows the basic texture to continue for a given amount of time. The unfolding of the piece, therefore, is entirely dependent upon the activity of the performer's brain. The idea here is that the performer should start in a calm state of mind which produces low frequencies, and then gradually increase the brain activity resulting in higher pitches, until the percussive sequence is triggered. In order to aid this control, the performer dances, at first making simple movements, but gradually increasing their complexity, which causes an increase in the frequency of the brain activity.

On the CD-ROM, in the folder **brainwav**, there is a demonstration version of the IBVA software for Macintosh platforms with examples. For more information please refer to the documentation provided.

8.15 Algorithmic but live: M

M is one of the first commercial programs for algorithmic composition to appear on the market: it was developed in the mid 1980s by David Zicarelli, Joe Chabade, John Offenbach and Antony Widoff. It was particularly popular with Amiga and Atari users; the latter was one of the best platforms for MIDI available at the time. Cycling74, best known for its fabulous Max programming system, has revived M, now in version 2.5, but only for the Macintosh platform (the original Atari version can still be found on the Internet). A demo version is available on the CD-ROM in folder **m**; the demo is fully functional except that the save function is disabled.

What is interesting about M is that it allows the computer to become a real partner during the compositional process. Basically, the user enters musical ideas and materials to be used in the piece, such as melodic phrases, pitch probabilities, rhythmic patterns, dynamic ranges, tempo ranges, and so forth, and then M lets you control how these process will be in real time; there is a Conducting Grid which allows the user to affect the operation of the entire program during the generative process. M sports a variety of mechanisms that let users work so interactively with the computer, that the distinction between composing and performing becomes completely blurred. The user can assign MIDI notes to specific functions that control the program, making M excellent software to use in conjunction with alternative MIDI controllers and interfaces such as the IBVA system introduced earlier.

Epilogue

The computer is undoubtedly a powerful tool for musical composition: it enables composers to try out new musical systems, to build rules-based generative programs, to map extra-musical formalisms onto musical representations, to build surrogate worlds for evolving musical cultures, and so forth. The range of possibilities is overwhelming.

This book emphasises a particular way of using the computer: as a generative tool. It introduces a number of generative techniques ranging from musical grammars and probabilistic tables, to fractals, chaos and evolutionary approaches. It also introduces the basics of machine learning of musical behaviour, with a view to embedding learning mechanisms into compositional systems. There has been a growing interest, however, in using the computer for composing and interacting with performers in real-time on stage. Such systems range from automatic accompaniment and karaoke-like systems, to methods for making the computer listen and react according to pre-defined compositional strategies. Apart from the discussion on controlling music using brainwaves, this book does not explicitly touch on the subject of interactive music. Although the great majority of composers today firmly purport the notion that composition is essentially a non-real-time activity that involves careful thinking, planning and crafting, interactive music composition is a considerable topic that deserves a book in its own right. M, IBVA and, to a certain extent, Texture and Vox Populi, are the only genuine systems on the CD-ROM designed primarily for real-time music making. Nevertheless, most of the techniques discussed in this book could well be applied to interactive musical systems. A good introduction to interactive music systems can be found in the book *Interactive Music Systems*, by Robert Rowe (1993). Also a couple of chapters on this topic can be found in the book *Readings in Music and Artificial Intelligence* edited by this author for Harwood Academic Publishers (Miranda, 2000).

In addition to focusing on the role of the computer as a generative tool, the present book also focuses on musical composition at the note level. For historical reasons (standard notation, for example), this level still provides the most natural framework within which to introduce new compositional thought. It should not, however, be difficult to transpose most, if not all, of the paradigms discussed here to lower or higher levels of musical structure, from sound synthesis to abstract manipulation of musical form.

By way of a conclusion, one of the lessons that this author has learned the hard way is that computers are extremely good at complying with systematisations and rules, but they are useless at breaking them. The Brazilian composer and teacher Frederico Richter (Frerídio), once made a profound comment that fits well in the context of this book: 'In music, rules are made to be broken. Good composers are those who manage to break them well.' In Chapter 7, a fairly sophisticated combinatorial procedure coupled with moulding rules is used to produce a short musical clip. In a sense the clip is formally consistent and looks beautiful on paper, but it does not sound quite right. A skilled composer would in a few minutes, however, be able fix this clip by editing a few notes, deleting some and adding others here and there, thus breaking the inherent rules of the system. This is an extremely difficult, if not impossible, skill to embody in a computer program, at least with technology as we know it today. Computers can indeed be programmed to compose music, but the truth is that they will seldom produce interesting music without human intervention at some point during or after the generative process. This leads us to the questions of what is 'interesting music' and what did Professor Richter mean by 'breaking the rules well'? Well, this book cannot offer definite answers to these questions: only your ears can tell...

Appendix 1: Excerpt from J. S. Bach's Chorale BWV 668

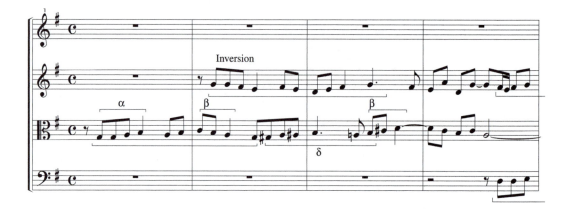

Figure Appendix 1 An excerpt from a Bach chorale featuring partial self-similar structures. The three lowest voices imitate the chorale in diminution and the alto in inversion. Brackets indicate other relationships. Score and annotations kindly provided by composer Gerald Bennett from the Swiss Center for Computer Music.

Appendix 2: Musical clip

Figure Appendix 2 The musical clip generated from the application of the moulding rules to chord sequences, as described in Chapter 7. The outcome from the strict application of rules in music composition seldom sounds absolutely right and this example is no exception. Nevertheless, with a few edits here and there, plus adequate instrumentation for performance, this clip can sound quite convincing.

Appendix 3: Formant chart

	/a/ M	/a/ F	/e/ M	/e/ F	/i/ M	/i/ F	/o/ M	/o/ F	/u/ M	/u/ F
F_1 Hz	622.25	783.99	392	392	261.63	349.23	392	440	349.23	329.63
Bwd_1 Hz	60	80	40	60	60	50	40	70	40	50
Amp_1 dB	90	90	90	90	90	90	90	90	90	90
F_2 Hz	1046.5	1174.7	1661.2	1568	1760	1661.2	783.99	783.99	587.33	739.99
Bwd_2 Hz	70	90	80	80	90	100	80	80	80	60
Amp_2 dB	83	86	78	66	60	70	79	81	70	78
F_3 Hz	2489	2793.8	2489	2793.8	2489	2793.8	2489	2793.8	2489	2793.8
Bwd_3 Hz	110	120	100	120	100	120	100	100	100	170
Amp_3 dB	81	70	81	60	74	60	69	74	58	60

Figure Appendix 3 Formant values for five vowels /a/ (as in the word 'car' in English), /e/ (as in the word 'bed'), /i/ (as in the word 'it'), /o/ (as in the word 'hot') and /u/ (as in the word 'rule') produced by a male and a female voice.

Appendix 4: A primer in Lisp programming

Lisp presents itself to the user as an interpreter; it works both as a programming language and as an interactive system: it waits for some input or command from the user, executes it, and waits again for further input. Lisp is an acronym for List Processor; almost everything in the Lisp world is a list. In programming terms, a list is a set of elements enclosed between parentheses, separated by spaces. The elements of a list can be numbers, symbols, other lists, and indeed, programs. Examples of lists are:

```
(the number of the beast)
(1 2 3 5 8 13)
(0.666 (21 xyz) abc)
(defun plus (a b) (+ a b))
```

Lisp is excellent at manipulating lists; there are functions that can do almost any operation you can possibly imagine with lists. Programs are lists themselves and they have the same simple syntax of lists. The following example illustrates what happens if one types in a short program for the Lisp interpreter. The sign '>' is the command line prompt indicating that the interpreter is waiting for your command (this prompt may vary from implementation to implementation):

```
> (+ 330 336)
666
```

When the interpreter computes (or *evaluates*, in Lisp jargon) a list, it always assumes that the first element of the list is the name of a function and the rest of the elements are the arguments that the function needs for processing. In the above example, the interpreter performs the addition of two numbers; the name of the function is the symbol '+' and the two arguments are the numbers 330 and 336.

If an argument for a function is also a function itself, then the argument is evaluated first, unless otherwise specified. The general rule is that the innermost function gets evaluated first. In the case of various nested functions, the outermost one is normally the last one to get evaluated. The Lisp interpreter treats each nested list or functions independently. For example:

```
> (+ 330 (* 56 (+ 5 1)))
666
```

In this example, the function '+' is evoked with number 330 as the first argument and with the result of the evaluation of (* 56 (+ 5 1)) as the second argument, which in turn evokes the function '*' with number 56 as the first argument and with the result of the evaluation of (+ 5 1) as the second argument. The result of the innermost step is 6, which is then multiplied by 56. The result of the multiplication is 336, which is ultimately added to 330. The result of the whole function therefore is 666.

Lisp programs often involve a large number of nested functions and data. In order to keep the code visually tidy, programmers normally break the lines and tabulate the nesting. For example:

```
(+ 330
   (* 56
       (+ 5 1)
   )
)
```

In order to write Lisp functions and programs, the language provides an instruction (or *macro*, in Lisp parlance), named *defun*; short for define function. The following example illustrates the definition of a new function labelled as *plus*:

```
> (defun plus (a b) (+ a b))
PLUS
```

The first argument for *defun* (i.e., second element of the list) is the name of the new function (e.g., *plus*) followed by a list of arguments that the new function should receive when it is called up (e.g., *a* and *b*). The last argument for *defun* is the body of the new function specifying what it should actually do with the arguments. The new *plus* function defined above executes the mere sum of two arguments *a* and *b*, but it could become a very complex algorithm or an entire program.

A variable is a symbol that represents the memory location for storing information. Variables may be used in Lisp to represent various types of information, such as a number, a list, a program, a single sound sample, a stream of samples or an entire synthesis instrument. The instruction normally used to allocate variables is *setf*. The first argument for *setf* is the label of the variable, and the second argument is the information to be associated to the variable. The following example creates a variable called *series* and associates the list (330 336 666) to this new variable:

> (setf series '(330 336 666))
(330 336 666)

In Lisp, an inverted comma before a list indicates that its first element is merely a unit of an ordinary list of data, and not a command. From now on, the variable *series* can be used in place of the list (330 336 666). For example, assume that the function *first* outputs the first element of a given list:

> (first series)
330

References

Abrams, S., Fuhrer, R., Oppenheim, D., Pazel, D. and Wright, J. (2000). A Framework for Representing and Manipulating Tonal Music. *Proceedings of the International Computer Music Conference*, San Francisco (USA): ICMA.

Abrams, S., Oppenheim, D., Pazel, D. and Wright, J. (1999). Higher-level Composition Control in Music Sketcher: Modifiers and Smart Harmony, *Proceedings of the International Computer Music Conference*, San Francisco (USA): ICMA.

Anderson, D.P. and Kuivila, R. (1992). Formula: A Programming Language for Expressive Computer Music. In: *Readings in Computer-Generated Music*, D. Baggi (ed.). Los Alamitos (USA): IEEE Computer Society Press, pp. 9–23.

Baggi, D.L. (1992). NeurSwing: An intelligent workbench for the investigation of swing in jazz. In: *Readings in Computer-Generated Music*, D. Baggi (ed.). Los Alamitos (USA): IEEE Computer Society Press.

Barnsley, M. (1988). *Fractals Everywhere*. London (UK): Academic Press.

Barrière, J.-B. (ed.) (1991). *Le timbre, métaphore pour la composition*. Paris (France): Christian Bourgois/Ircam.

Bel, B. (1998). Migrating Musical Concepts: An Overview of the Bol Processor, *Computer Music Journal*, **22**, No. 2, pp. 56–64.

Bernstein, L. (1976). *The Unanswered Question*. Cambridge (USA): Harvard University Press.

Best, C.T. (1988). The emergence of cerebral asymmetries in early human development: A literature review and a neuroembryological model. In: *Brain Lateralization in Children*, D.L. Molfese and S.J. Segalowitz (eds). New York (USA): The Guildford Press.

Bharucha, J.J. and Olney, K.L. (1989). Tonal cognition, artificial intelligence and neural nets. *Contemporary Music Review*, **4**, 341–56.

Biles, J.A. (1994). GenJam: A Genetic Algorithm for Generating Jazz Solos. *Proceedings of the International Computer Music Conference*, San Francisco (USA): ICMA.

Bilotta, E., Pantano, P. and Talarico, V. (2000). Synthetic Harmonies: an approach to musical semiosis by means of cellular automata. *Artificial Life VII: Proceedings of the Seventh International Conference on Artificial Life*, M.A. Bedu *et al.* (eds). The MIT Press.

Boulez, P. (1963). *Penser la musique aujourd'hui*. Geneva (Switzerland): Gonthier.

Brown, S. (2000). The Musilanguage Model of Music Evolution. In: *The Origins of Music*, N.L. Walling, B. Merker and S. Brown (eds). Cambridge (USA): The MIT Press.

Campbell, M. and Greated, C. (1987). *The Musicians Guide to Acoustics*. London (UK): J.M. Dent & Sons.

Cardew, C. (1974). *Stockhausen Serves Imperialism*. London (UK): Latimer New Dimensions.

Carpenter, R.H.S. (1990). *Neurophysiology*. London (UK): Edward Arnold.

Carpinteiro, O. (1995). A neural model to segment musical pieces, *Proceedings of the Second Brazilian Symposium on Computer Music*, pp. 114–20, Porto Alegre (Brazil): SBC.

Chomsky, N. (1957). *Syntactic Structures*. The Hague: Mouton and Co.

Cood, E.F. (1968). *Cellular Automata*. London (UK): Academic Press.

Cook, N. (1987). *A Guide to Musical Analysis*. London (UK): J.M. Dent & Sons.

Cope, D. (1987). An Expert System for Computer-Assisted Composition. *Computer Music Journal*, **11**, No. 4, 30–40.

Cope, D. (1991). *Computers and Musical Style*. Oxford (UK): Oxford University Press.

Dannenberg, R. (1997). Machine Tongues XIX: Nyquist, a Language for Composition and Sound Synthesis. *Computer Music Journal*, **21**, No. 3, 50–60.

Darwin, C. (1859). *On the origins of species by means of natural selection or the preservation of favoured races in the struggle for life*. London (UK): Murray.

Dawkins, R. (1986). *The Blind Watchmaker*. London (UK): Penguin Books.

Degazio, B. (1997). The Evolution of Musical Organisms. *Leonardo Music Journal*, **7**, 27–33.

Despins, J.-P. (1996). *La Música y el cerebro*. Barcelona (Spain): Editorial Edisa.

Dewdney, A.K. (1989). A cellular universe of debris, droplets, defects and demons. *Scientific American*, **261**, No. 2, 102–5.

Dolson, M. (1989). Machine Tongues XII: Neural networks. *Computer Music Journal*, **13**, No. 3, 28–40.

Ekerland, I. (1995). *Le Chaos*. Paris: Flammarion.

Epstein, J.M. and Axtell, R. (1996). *Growing Artificial Societies: Social Sciences from the Bottom Up*. Cambridge (USA): The MIT Press.

Ermentrout, G.B. and Edelstein-Keshet, L. (1993). Cellular Automata Approaches to Biological Modeling. *Journal of Theoretical Biology*, **160**, 97–133.

Freisleben, B. (1992). The neural composer: A network for musical applications. *Proceedings of the International Conference on Artificial Neural Networks*, Vol. 2, pp. 1663–6, Amsterdam (The Netherlands): Elsevier.

Gilbert, G.N. and Troitzsch, K.G. (1999). *Simulations for the Social Scientist*. Buckingham (UK): Open University Press.

Grimmet, G.R. and Stirzaker, D.R. (1982). *Probability and Random Processes*. Oxford (UK): Oxford University Press.

Helmholtz, H. (1954). *On the Sensations of Tone*. New York (USA): Dover.

Hofstadter, D.R. (1979). *Gödel, Escher, Bach: An Eternal Golden Braid*. London (UK): The Harvester Press.

Hogeweg, P. (1988). Cellular Automata as a Paradigm for Ecological Modeling. *Applied Mathematics and Computation*, **27**, 81–100.

Holtzman, S.R. (1994). *Digital Mantras: The Languages of Abstract and Virtual Worlds*. Cambridge (USA): The MIT Press.

Howard, D.M. and Angus, J. (1996). *Acoustics and Psychoacoustics*. Oxford (UK): Focal Press.

Hush, D.R. and Horne, B.G. (1993). Progress in Supervised Neural Networks. *IEEE Signal Processing Magazine*, **10**, No. 1, 8–39.

Jacob, B.L. (1995). Composing with Genetic Algorithms. *Proceedings of the International Computer Music Conference*, pp. 452–5, San Francisco (USA).

James, J. (1993). *The Music of the Spheres: Music, Science and the Natural Order of the Universe*. London (UK): Abacus.

Jan, S. (2000). Replicating Sonorities: Towards a Memetics of Music. *Journal of Memetics – Evolutionary Models of Information Transformation*, **1**, No. 1. (http://www.cpm.mmu.ac.uk/jom-emit/2000/vol4/index.html)

Janata, P. (1995). ERP measures assay the degree of expectancy violation of harmonic contexts in music. *Cognitive Neuroscience*, **7**, No. 2, 152–64.

Jones, D. (1956). *An Outline of English Phonetics*. Cambridge (UK): Heffer.

Jordan, R. and Kafalenos, E. (1994). Listening to Music: Semiotic and Narratological Models. *Musikometrika*, Vol. 6, M.G. Boroda (ed.). Bochum (Germany): Brockmeyer.

Jung, T., Makeig, S. and Stensmo, M. (1997). Estimating alertness from the EEG Power Spectrum. *IEEE Transactions on Biomedical Engineering*, **44**, 60–69.

Jusczyk, P.W. (1997). *The discovery of spoken language*. Cambridge (USA): The MIT Press.

Kippen, J. and Bel, B. (1992). Modelling Music with Grammars: Formal Language Representation in the Bol Processor. In: *Computer Representations and Models in Music*, A. Marsden and A. Pople (eds), London (UK): Academic Press.

Ladefoged, P. and Maddieson, I. (1996). *The Sounds of the World's Languages*. Oxford (UK): Blackwell.

Laine, P. (1997). Generating musical patterns using mutually inhibited artificial neurons, *Proceedings of the International Computer Music Conference*, pp. 422–5, San Francisco (USA): ICMA.

Langton, C.G. (ed.) (1997). *Artificial Life: an Overview*. Cambridge (USA): The MIT Press.

Larivaille, P. (1974). L'Analyse (morpho)logique du récit. *Poétique*, **19**, 368–88.

Lerdhal, F. and Jackendoff, R. (1983). *A Generative Theory of Tonal Music*. Cambridge (USA): The MIT Press.

Lieberman, P. (1998). *Eve Spoke*. London (UK): Picador/Macmillan.

Linster, C. (1990). A neural network that learns to play in different styles, *Proceedings of the International Computer Music Conference*, pp. 311–13, San Francisco (USA): ICMA.

Lipmann, R.P. (1987). An introduction to computing with neural nets, *IEEE Acoustics, Speech and Signal Processing Magazine*, **4**, No. 2, 4–22.

Loy, D.G. (1991). Connectionism and Musiconomy. *Music and Connectionism*, P.M Todd and D.G. Loy (eds), Cambridge (USA): MIT Press.

Maconie, R. (ed.) (1991). *Stockhausen on Music*. London (UK): Marion Boyars.

McAdams, A. (ed.) (1987) Music and psychology: a mutual regard. *Contemporary Music Review*, Vol. 2, Part 1, Reading (UK): Gordon & Breach.

McAlpine, K. (1999). Applications of Dynamical Systems to Music Composition, PhD Thesis, Department of Mathematics, University of Glasgow.

Malt, M. (1999). Reflexiones sobre el acto de la composición. *Música y Nuevas Tecnologías: Perspectival para el Siglo XXI*, E. Miranda (ed.), Barcelona (Spain): L'Angelot.

Mandelbrot, B. (1982). *The Fractal Geometry of Nature*. New York (USA): W.H. Freeman.

Manning, P. (1985). *Electronic and Computer Music*. Oxford (UK): Oxford University Press.

Meyer, L. (1956). *Emotion and Meaning in Music*. Chicago (USA): Chicago University Press.

Miranda, E.R. (1998). *Computer Sound Synthesis for the Electronic Musician*. Oxford (UK): Focal Press.

Miranda, E.R. (2000). (ed.) *Readings in Music and Artificial Intelligence*. Amsterdam (The Netherlands): Harwood Academic Publishers.

Miranda, E.R. (2001). *Entre o Absurdo e o Mistério*. (Musical score for chamber orchestra with parts.) Porto Alegre (Brazil): Edições Musicais Goldberg.

Nelson, G.L. (1995). Sonomorphs: An Application of Genetic Algorithms to the Growth and Development of Musical Organisms, unpublished paper, Oberlin (USA): Conservatory of Music.

Oppenheim, D.V. (1994) Slappability: A New Metaphor for Human Computer Interaction. In: *Music Education: An Artificial Intelligence Approach*, M. Smith *et al.* (eds), London: Springer Verlag, pp. 92–107.

Orton, R. (1990). Music Fundamentals, unpublished lecture notes, Department of Music, University of York.

Orton, R. and Kirk, P.R. (1992). Tabula Vigilans, *Proceedings of the International Computer Music Conference*, San Francisco (USA): ICMA.

Pask, G. (1961). *An Approach to Cybernetics*. London (UK): Hutchinson.

Preston, K. and Duff, M.J.B. (1984). *Modern Cellular Automata: Theory and Applications*. New York (USA): Plenum Press.

Rayward-Smith, V.J. (1983). *A First Course in Formal Language Theory*. Oxford (UK): Blackwell.

Reck, D. (1997). *Music of the whole earth*. New York (USA): Da Capo Press.

Revill, D. (1992). *The roaring silence*. London (UK): Bloomsbury.

Richman, B. (2000). How Music Fixed Nonsense into Significant Formulas: On Rhythm, Repetition and Meaning. In: *The Origins of Music*, N.L. Walling, B. Merker and S. Brown (eds), Cambridge (USA): The MIT Press.

Rosenblatt, F. (1958). The perceptron: A probabilistic model for information storage and organisation of the brain. *Psychological Review*, **65**, 368–408.

Rosenboom, D. (1990). Extended Musical Interface with the Human Nervous System. *Leonardo Monograph No. 1*. Cambridge (MA): The MIT Press.

Rossing, T.D. (1990). *The Science of Sound*. Reading (USA): Addison-Wesley.

Rowe, R. (1993). *Interactive Music Systems*. Cambridge (USA): The MIT Press.

Rumsey, F. (1994). *MIDI Systems and Control*. Oxford (UK): Focal Press.

Saiwaki, N., Kato, K. and Inokushi, S. (1997). An Approach to Analysis of EEGs Recorded During Music Listening. *Journal of New Music Research*, **26**, pp. 227–43.

Schoenberg, A. (1967). *Fundamentals of Musical Composition*. London (UK): Faber and Faber.

Schwanauer, S.M. and Levitt, D.A. (eds) (1993). *Machine Models of Music*. Cambridge (USA): The MIT Press.

Smith, P. (1998). *Explaining Chaos*. Cambridge (UK): Cambridge University Press.

Stauffer, D. and Stanley, H.E. (1990). *From Newton to Mandelbrot: A Primer in Theoretical Physics*. London (UK): Springer-Verlag.

Steels, L. (1997). The Synthetic Modeling of Language Origins, *Evolution of Communication Journal*, **1**, No. 1, 1–34.

Stockhausen, K. (1991). Four Criteria of Electronic Music. In: *Stockhausen on Music*, R. Maconie (ed.), London (UK): Marion Boyars, pp. 88–111.

Storr, A. (1993). *Music and the Mind*. London (UK): HarperCollins.

Sundberg, J. and Lindblom, B. (1993). Generative Theories in Language and Music Descriptions. In: *Machine Models of Music*, S.M. Schwanauer and D.A. Levit (eds), Cambridge (USA): The MIT Press.

Sutherland, R. (1994). *New Perspectives in Music*. London (UK): Sun Tavern Fields.

Swade, D. (1991). *Charles Babbage and his Calculating Engines*. London (UK): Science Museum.

Swingler, K. (1996). *Applying Neural Networks: A Practical Guide*. London (UK): Academic Press.

Taube, H. (1997). An Introduction to Common Music. *Computer Music Journal*, **21**, No. 1, 29–34.

Thomas, D.A. (1995). *Music and the origins of language*. Cambridge (UK): Cambridge University Press.

Todd. P. (1989). A Connectionist Approach to Algorithmic Composition. *Computer Music Journal*, **13**, No. 4, 27–43.

Todd, P. (2000). Simulating the Evolution of Musical Behaviour. In: *The Origins of Music*, N.L. Walling, B. Merker and S. Brown (eds), Cambridge (USA): The MIT Press.

Toiviainen, P. (2000). Symbolic AI versus Connectionism in Music Research. In: *Readings in Music and Artificial Intelligence*, E.R. Miranda (ed.), Amsterdam (The Netherlands): Harwood Academic Publishers.

Vince, A.J. and Morris, C.A.N. (1990). *Discrete Mathematics for Computing*. London (UK): Ellis Horwood.

Washabaugh, W. (1995). The Politics of Passion: Flamenco, Power, and the Body. *Journal of Musicological Research*, **15**, 85–112.

Weng, W. and Khorasani, K. (1996). An Adaptive Structure Neural Networks with Application to EEG Automatic Seizure Detection. *Neural Networks*, **9**, No. 7, 1223–40.

Wiener, N. (1948). *Cybernetics*. Cambridge (USA): MIT Press and John Wiley.

Wittgenstein, L. (1963). *Philosophical Investigations*. Oxford (UK): Basil Blackwell.

Xenakis, I. (1967). *Herma* (musical score). London (UK): Boosey & Hawkes Music Publishers.

Xenakis, I. (1971). *Formalized Music*. Bloomington (USA): Indiana University Press.

CD-ROM instructions

==

In order to run or install a program on your computer, you should normally copy the respective folder onto your hard disk and follow the usual procedures for running or installing software. Most packages provide a Readme file with installation instructions; use Notepad (on PC) or SimpleText (on Mac) to read a Readme file. A few general tips for each program are given below. It is strongly advised that you read them before you go on to use the programs.

Some packages provide HTML documentation. In principle you should be able to read them with any browser. They have been kept as simple as possible, stripped of fancy frames, backgrounds and tricky links. The objective is to convey the information in a problem-free manner. A few packages provide documentation in PDF format. In order to read these files you will need to install Adobe Acrobat Reader; this is freely available from Adobe's web site: <http://www.adobe.com>.

Note that folders, file names and web addresses are written between '<' and '>'.

The folder <various> contain miscellaneous materials other than software:

(a) The folder <midi> contains MIDI files, kindly provided by James Correa, for the various musical examples given in the book. They are labelled with the number of the relevant figure. These files are set to play GM MIDI 01, piano, by default; even if they are polyphonic examples. In any case, you can load the MIDI files onto your favourite MIDI sequencing program and specify whatever timbres you wish.

(b) The folder <evolution> contains an HTML document with an animation that complements the notions introduced in Chapter 6. Also, there are two voice synthesisers as discussed in Chapters 6 and 7: <singer.txt> and <articulator.txt>. The code contains some instructions on running them. Refer to Nyquist documentation.

(c) The folder <weblinks> contains an HTML document with URL links to the people and institutions that supplied the materials for the CD-ROM and other related links.

(d) The folder <compositions> contains MP3 files (and WAV for PC or AIFF for Mac) which are excerpts from two compositions, plus the score for one of them, in PDF and EPS formats. Files <reck1.mp3> and <reck2.mp3> are excerpts from the piece 'Entre o Absurdo e o Mistério', for chamber orchestra (recording of its premiere in Edinburgh, UK) and the file <vanpoint.mp3> is the third movement of the piece 'Grain Streams' for piano and electroacoustics (recording of its premiere in Annecy, France). Both pieces were composed with the aid of CAMUS (discussed in Chapter 6). Readers will need an MP3 player to listen to the compositions and a suitable PDF or EPS reader to see and/or print the score; e.g. Adobe Acrobat (PDF) or Aladdin Ghostscript (EPS): <http://www.aladdin.com>. The full score and parts for 'Entre o Absurdo e o Mistério' is available from: Goldberg Edicoes Musicais Ltda, Rua Sao Vicente, 546/109, 90630-180, Porto Alegre, RS, Brazil; Tel/Fax: +55 51 333 4101, Email: <edgberg@nutecnet.com.br>.

PC software

Tested on a Sony VAIO laptop, Pentium MMX 300 MHz, running Windows 98, with an Ego-Sys PCMCIA WaMiBox sound/MIDI interface.

Also tested on a Pentium 166 MHz machine, 32 Mb RAM, running Windows 95 with SoundBlaster AWE64 Gold card and on a 500 MHz machine, 128 Mb RAM, running Windows 98 with an Emagic Audiowerk sound card.

1 CAMUS and CAMUS 3D <camus>

CAMUS and CAMUS 3D are two different programs; they are in <camus\camus> and <camus\camus3d> respectively. Copy the main <camus> folder onto your hard disk, double-click on <Setup.exe> and follow the respective procedures on the screen.

The on-line Help of the programs is perhaps the best way to get started. The tutorial in the folder <tutorial> is very comprehensive and it is an excellent complement to the theory introduced in the book: <camus.htm>. Both CAMUS and CAMUS 3D are freeware.

2 FractMus <fractmus>

Copy the <fractmus> folder onto your hard disk, double-click on <Setup.exe> and follow the procedures. The program is straightforward to operate and the on-line Help shows how to get started. There is also a good introduction to the various fractal algorithms featured in the program, which makes a superb complement to the book. A number of examples are given, including John Cage's 4'33". FractMus is freeware.

3 Harmony Seeker <harmseek>

In order to install the system, firstly uncompress the <harmseek.zip> ZIP folder onto your hard disk.

Harmony Seeker is programmed in Java; it needs the Java Runtime Environment (JRE) to run (basically, JRE will simulate Java computer on your PC). If you do not have JRE on your machine, you will have to install it first. In order to install JRE, simply double-click on <Setupjre.exe> in the folder <...\install> and follow the instructions. Then double-click on <setup.exe> in order to install Harmony Seeker.

Harmony Seeker is still being developed; the authors have kindly put together a prototype especially for this book. There is a user guide in PDF format that explains what it can do at the moment and how to get started: <QuickStart.pdf>.

Be aware that when you create a new project as it says in the user guide, the new project window is not activated by default; its title bar is coloured grey. You have to activate the window before you press the Play button to run the program, otherwise nothing will happen. The activation of the window is done by clicking on its title bar: it should then become blue or purple.

4 The Koan suite <koanpc>

Koan X demo, Koan Pro demo, Koan file player and Koan plugin for PC are available in folder <koanpc>. Each executable in this folder corresponds to one of these four programs; they have easily identifiable labels. In order to install a program simply run the respective executable and follow the instructions on the screen. The documentation is professional and the programs feature a number of examples.

The version of the Koan suite provided on the CD-ROM is as of December 2000. SSEYO strongly advise readers to check SSEYO's web site regularly for new updates: <http://www.sseyo.com>.

5 A Music Generator <musigen>

In order to install a Music Generator, copy the <musigen> folder onto your hard disk, double-click on <Setup.exe> and follow the procedures. There is a program preview in HTML format in folder <musigen\preview>. This program is straightforward enough to get by, but a better understanding of the basic concepts behind its functioning will help you to get most of it. There is a good tutorial in the on-line Help documentation which is a must to get started.

A Music Generator is shareware and costs $25. You can try it out for 30 days. If you decide to keep using it then you are required to register the program; refer to instructions in the on-line Help. As a registered user you are entitled to technical support and to eventual upgrades.

6 Nyquist <nyquistpc>

This is a fully fledged freeware sound synthesis and composition programming language. Nyquist is a very powerful tool for composition as it combines the features of synthesis-only languages (such as Csound) with composition-oriented capabilities. Nyquist does not need installation as such; simply copy the whole <nyquistpc> folder onto your main hard disk.

You will find the executable <nyquist.exe> in the <runtime> folder. Run this program and you will see a text window with some indication that Nyquist has started and has loaded some files. At this point the Nyquist is ready to take in Nyquist or Lisp expression such as the ones given in the examples given in section 1.3 of the user's manual: <nyqman.pdf>.

Once you get ready to do serious work with Nyquist, you will probably want to keep your projects and composition separate from the <runtime> folder where Nyquist is presently located; <introduction.html> explains how to set this up.

There are additional tutorials and documentation, kindly provided by Pedro J. Morales, in folder <nyquistpc\demos>. Here you will find a number of examples on how to explore the potential of Nyquist. Also, a number of Nyquist programs are available in the folder <nyquist\test>.

7 Vox Populi <populi>

Vox Populi is freeware and can be found in the folder <populi>. In order to install Vox Populi, copy this folder onto your hard disk, run <Setup.exe> and follow the instructions on the screen. A short user's guide is provided in Readme.txt and a tutorial in HTML format can be found in the folder <populi\voxtutor>. This tutorial will introduce you to the underlying concepts of Vox Populi's generative engine.

If, during installation, you receive an error message saying 'COMMOLG.DLL is in use. Please close all applications and re-attempt Setup' just ignore it. Press the Ignore button and the setup should resume with successful installation.

8 Music Sketcher <sketcher>

The installer for Music Sketcher can be found in the folder <sketcher>. Simply run the executable <setupex.exe> and follow the instructions on the screen. The <Readme.txt> contains important information on how to get started and the on-line Help menu contains all the information you need to operate the program. Also, a number of interesting examples and riffs are provided: check folders <..\Samples> and <..\Content>. Music Sketcher's inventors wish to emphasise that this program is not an IBM product, ready for the market; it is rather a prototype that incorporates some aspects of their research work. Nevertheless, the program is robust, functions well and can produce interesting results. This version on the CD-ROM is fully functional and it is freeware.

9 Tangent <tangent>

Tangent is available in the folder <tangent>. Copy this folder onto a temporary directory of your hard disk and run its installer <setup.exe>, which can be found in the folder <program>.

The best way to get started in Tangent is to read the introductory document <thelp10.rtf>. This document is saved in RTF format but most word processors should be able to open it. A reference manual, also in RTF, is available; note, however, that the installer also places an MS Word DOC version of the manual in Tangent's directory.

There are a number of impressive examples available. Note that once you have loaded a demo file (by pressing the button LoadComp), you should then render it (button Composer) in order to playback.

Tanget is shareware. Please consult the documentation for more information on how to register.

10 Texture <texture>

Texture, in folder <texture>, is freeware and it does not require installation as such. Simply copy the whole folder onto your hard disk. The folder <texture\doc> contains various MS Word documents, each of which focuses on a determined aspect of Texture. RTF and TXT versions of these documents are also available in <texture\doc\rtf> and <texture\doc\txt> respectively. A variety of examples can be found in <texture\examples>.

11 Distance Networks <camus\disnet>

Kenny McAlpine's neural network designed to provide a measure of how far apart two notes are harmonically is available in the folder <camus\disnet>. Run the executable <install.exe> to install the program. Instructions on how to use it can be found in the <ReadMe.txt> file.

Macintosh software

Tested on a Macintosh G3 Blue, 350 MHz under Mac 8.5 and OS 9.0.4, with an Audiomedia III card and Opcode Midiport 32.

12 OpenMusic <openmusic>

OpenMusic needs LISP (Common Lisp) to function. The sources are available (they can be found on the CD-ROM), but you will need a LISP compiler (more specifically, the Digitool MCL compiler) to compile it. A compiled, ready to run demo version of OpenMusic is also available on the CD-ROM, in folder <openmusic:OM3.5demo>. The limitation of this demo version is that it has been compiled with a demo version of the Digitool MCL compiler which lets the program run for 15 minutes only.

In order to run the demo version, copy the folder <OM3.5demo> (or expand the <OM3.5demo.sea> file) onto your hard disk and follow the instructions in <README>; the program is in the <openmusic:OM3.5demo:Image folder>. The HTML documentation in the folder <OMUserDocumentation> provides an excellent introduction to the system.

The sources for compiling <OpenMusic> are compressed in <OM3.5GPLRelease.sea.bin> and tutorial codes can be found in <OMTutorialCode.sea>. If you are interested in accessing these, just expand them onto your hard disk accordingly. Refer to the <README.txt> document for more information.

13 Koan Pro and Koan Plugin <koanmac>

Koan Pro and Koan Plugin for the Macintosh platform are available in the folder <koanmac>.

Koan Pro (in the folder <koanmac:KProDemo>) is a demo version. Please read the <Read Me> file before running the program. The <koanpro_help.htm> document provides an excellent insight into the system. The full version of Koan Plugin for the Internet is available in the folder <koanmac:koanplugin>. In order to install it just double-click on the <InstallKoanPlugin>

icon. Please refer to the <rkplg.txt> document for more information.

The SIT files on the CD-ROM are compressed versions of the two packages.

The version of the Koan suite provided on the CDROM is as of December 2000. SSEYO strongly advise readers to check SSEYO's web site regularly for new updates: <http://www.sseyo.com>.

Make sure that you have OMS installed, otherwise Koan Pro will not launch.

14 IBVA <ibva>

A demo of the IBVA system is available in the <ibva> folder. In fact there are two demo programs here: the IBVA itself and Alps+; the latter is used to map brainwaves signals captured by IBVA onto a MIDI unit. In order to install the system, simply copy the entire <ibva> folder onto the root of your startup disk. The system is composed of a number of applications. The executable <runALL> launches a demo involving all these applications. This activates a script with a series of commands for the Mac OS that automatically run IBVA, Alps+, etc. This is perhaps the best way to get to know the system but WARNING: before you run <runALL> you must rename the <ibva> folder that you copied onto the root of your startup hard disk to <IBVAmusic>. (The script will look for the applications in this directory to launch them.). You need the MIDI Manager properly installed in order to be able to hear the music. Please refer to the <Read Me> file for more information. IBVA's manual is also presented in the form of a Mac script; double-click on the icon <IBVA2p Manual> in order to open it.

15 Nyquist <nyquimac>

This is a fully fledged sound synthesis and composition programming language. Nyquist is a very powerful tool for composition as it combines the features of synthesis-only languages (such as Csound) with composition-oriented capabilities. Nyquist does not need installation as such; simply copy the whole <nyquimac> folder onto your hard disk.

The executable is in the folder <nyquimac:Nyquist:runtime>. Please read both <README> files sitting in the folders <nyquist> and <nyquimac:Nyquist>. The user manual is very comprehensive and you will probably refer to it extensively if you use Nyquist for your projects. The manual is provided in three formats: HTML, PDF and TXT. In folder <nyquimac:Nyquist:demos> there are additional tutorials and documentation specially prepared for this book. Here you will find many didactic examples to explore the potential of Nyquist. Also, a number of Nyquist programs are available in the folder <nyquimac:Nyquist:test>.

16 Bol Processor <bolprocessor>

The full version of Bol Processor is available in the folder <bolprocessor>. The folder <BOL PROCESSOR BP2> (for which a SEA compressed version is also provided) contains a <Read first> file and the installer (<Install Bol Processor>). Read the instructions in the <Read first> document before installing the system.

Note that the BinHex files mentioned at the beginning of the <Read first> document have already been converted on the CD-ROM: these are the <bol-processor-doc> and <BOL PROCESSOR BP2> folders. The former contains the Bol Processor documentation in both MS Word and HTML formats. Also there is a lighter, but fairly complete, version of the documentation in the folder <bolprocessor:howard>. Note that installer will create two new folders in the directory selected for the installation on your hard disk: one called <bol-processor-application> and another <bol-processor-data>. Making changes in these folders is not recommended, but you may as well trash the <+sc.shutdown> script after running the program for the first time.

17 M <mdemo>

A demo version of M is available in folder <mdemo>. There is no need for installation: just copy this folder onto your hard disk. There is a nicely put together user's manual in PDF format containing all information you need to run the system. In the folder <mdemo:M Tutorial Files> there are four examples that you can load and explore. The demo version is fully functional, except that the saving function is disabled.

Index